The Backward Classes in Contemporary India

The Backward Classes in Contemporary India

ANDRÉ BÉTEILLE

DELHI
OXFORD UNIVERSITY PRESS
BOMBAY CALCUTTA MADRAS

Oxford University Press, Walton Street, Oxford OX2 6DP
Oxford New York Toronto
Delhi Bombay Calcutta Madras Karachi
Kuala Lumpur Singapore Hong Kong Tokyo
Nairobi Dar es Salaam Cape Town
Melbourne Auckland Madrid

and associates in
Berlin Ibadan

First published, 1992
Second impression, 1994

ISBN 0 19 563035 1

Typeset by Imprinter, C-79 Okhla Phase-I, New Delhi 110020
Printed at Ram Printograph (India), New Delhi 110020
and published by Neil O'Brien, Oxford University Press
YMCA Library Building, Jai Singh Road, New Delhi 110001

Contents

Preface

The work entitled *The Backward Classes and the New Social Order*, based on the Ambedkar Lectures delivered in the University of Bombay on 6 and 7 March 1980, was published by Oxford University Press in 1981. I take advantage of the occasion of its reprinting to add an essay on a related theme, 'Distributive Justice and Institutional Well-Being', which is the text of the V. T. Krishnamachari Lecture delivered in the Institute of Economic Growth on 12 November 1990. These constitute the two principal parts of the present work, to which I have appended a number of shorter pieces published in newspapers and, in one case, in a monthly magazine. The shorter pieces sought to present to a wider public arguments that were more elaborately developed in the longer ones.

There is now a very large literature on the Backward Classes to which I have contributed my own share during the last thirty years. What is presented here is only a part of my work on the subject in which questions of policy receive more direct attention; I have written elsewhere about other aspects of the Backward Classes. In my work as a whole I have always tried to stress the need for a differentiated approach to them, since I believe that what holds for the Scheduled Castes and Tribes does not necessarily hold for the Other Backward Classes. Some changes have naturally taken place in the last thirty years in the climate of opinion regarding a policy for the Backward Classes but my initial scepticism about the social benefits of reservation has grown stronger over the years. I believe that reservations in education and employment are on the whole harmful, and that their further extension is incompatible with our basic social and economic objectives. At the same time, political pressures in favour of reservation will continue in the name of equality and social justice.

The two principal parts stand by themselves and do not call for any additional explanation. The only point to which I would like to draw attention here is that the Ambedkar Lectures were written before the Mandal Commission submitted its Report whereas the V. T. Krishnamachari Lecture was written in the wake of the announcement in August 1990 that the government was going to implement with immediate effect the recommendations of the Commission in regard to job reservation.

The shorter pieces were written, all but one, as newspaper articles. The exception is the piece entitled, 'Reservations: The Problem' which was written as the 'poser' for the December 1981 issue of *Seminar*. The newspaper articles were written, with one exception, after the submission of the Report of the Mandal Commission. 'The Indian Road to Equality' was published in August 1982, shortly after Parliament adopted the recommendations of the Commission with unanimous acclaim. 'Caste and Politics' was published in September 1990 in the wake of the executive order by Mr V. P. Singh's government to implement job reservations for the Other Backward Classes. To these post-Mandal pieces I have added one which was published in August 1961, long before the Mandal Commission had been thought of; I am pleased to note that there too I had expressed grave doubts about the wisdom of job quotas for the Other Backward Classes.

I have left the pieces as they were when first published, making only a few verbal changes. Some titles altered by the newspaper editors have been restored. I ought to draw attention to one change, however. My article, 'Caste and Politics' in *The Times of India* of 11 September 1990 ended with the words, 'for caste has no function today except in politics'. I was talking of modern institutions and I thought that the context made that clear. Several persons have pointed out that the phrase was ambiguous and misleading. I have now altered the words a little to make my statement more clear, and, I hope, more acceptable.

As before, I would like to express my thanks to the University of Bombay for the invitation to deliver the Ambedkar Memorial Lectures. I am grateful also to the Institute of Economic Growth for the invitation to deliver the V. T. Krishnamachari Memorial Lecture. My thanks are due, finally, to the editors of *Seminar* magazine, *The Times of India*, *The Hindustan Times, The Hindu* and *The Indian Express* for offering me the hospitality of their columns from time to time.

André Béteille

The Backward Classes and the New Social Order

I

I would like to devote these lectures to the problem of the Backward Classes in the new social order. In electing to do so I have in mind the concern for the problem of the person whose memory we are here to honour as well as the intrinsic importance of the subject itself. It is a subject that has a practical and a theoretical side, and if I choose to dwell especially on the latter, it is in the belief that action can be fruitful only if it is informed by proper understanding. A proper understanding of the problem of the Backward Classes requires us to view it in several perspectives, notably those of the social sciences and of legal studies; for we are at every step confronted by the divergence between what exists as social reality and what ought to exist according to the laws we have created for ourselves.

Put in a somewhat different way, what I propose to do may be described as a sociological critique of the equality provisions in the Constitution of India. These provisions are both wide-ranging and varied. We cannot understand either their scope or their complexity in terms of purely formal principles. We can appreciate their nature and significance only by relating them to the historical background from which they have emerged and the social context to which they were designed to apply.

No society can move forward unless it sets for itself an ideal of achievement that is superior to the present reality. To this extent the design for living enshrined in a Constitution must rise above the social arrangements that exist on the ground. At the same time, it cannot afford to lose touch with the social facts as they are. For these facts are not only there, but they exercise constraints that cannot be wished out of existence. A Constitution may indicate the direction in which we are to move, but the social structure will decide how far we are able to move and at what pace.

A society has thus to be judged both for what it is and for what it

The B. R. Ambedkar Memorial Lectures delivered at the University of Bombay on 6 and 7 March 1980.

wishes to be. A written Constitution, and especially one that is written at a decisive turn in its history, has a certain significance as an expression of what a society seeks to achieve for itself. A very striking feature of our Constitution is its stress on equality. It is present in the Preamble; it is present in the part embodying the Fundamental Rights; and it is present in the part laying down the Directive Principles of State Policy. Legislative enactments and judicial pronouncements have during the last three decades reiterated this commitment over and over again.

Jurists have pointed out how we have gone further than most modern Constitutions, including the American, in inscribing the commitment to equality into ours. Thus, in speaking of the guarantee of equality, P. K. Tripathi says, 'But it must be appreciated that the scope of the guarantee in the Constitution of India extends far beyond either, or both, the English and the United States guarantees taken together.'[1] One has only to go through the record of debates in the Constituent Assembly or to examine the notes and memoranda prepared by members of the Assembly and by the Constitutional Adviser to see how strong the preoccupation with equality was among the makers of the Indian Constitution. This preoccupation was itself a part of a historical process that grew with the movement for freedom from colonial bondage.

Despite all this, our practice continues to be permeated by inequality in every sphere. The marks of inequality are visible in every form of collective life. Our rural and urban communities are divided and subdivided into groups and categories that are ranked in elaborate gradations. Distinctions among castes and among classes, though no longer upheld by the law, are taken into account everywhere. There are numerous barriers between the strata, and they are difficult to cross. The reality of rigid social stratification makes itself felt in the daily lives of the poor and the oppressed in general, and the Untouchables in particular.

The problem of the Backward Classes is, in its most general form, the problem of achieving equality in a world permeated by inequality. The significance of the category 'Backward Classes' lies not only in its size and extent, but also in the uniquely Indian way of defining its boundaries. This uniqueness is a reflection of specific social and historical conditions. In India, unlike in other societies, 'backwardness' is viewed as an attribute not of individuals but of communities which are, by their nature, self-perpetuating. In ordinary sociological discourse a class is a set of individuals—or, at best, families—sharing certain life chances in common that they may or may not owe to their ancestors, and that they may or may not transmit to their descendants. By the terms of that discourse, the Backward Classes are not classes at all, but groups of communities.

Judicial pronouncements on the subject reflect the ambiguity inherent in the situation. There are judgements, as in the famous *Balaji*'s case, which imply that there is, or ought to be, a clear distinction between 'caste' and 'class'.[2] There are other judgements which maintain, as in *Rajendran*'s case, that 'a caste is also a class of citizens'[3] or more strongly, as in *Periakaruppan*'s case, that 'A caste has always been recognized as a class.'[4] The discrepancy between the two views is in part, but only in part, due to the use in the first case of a 'sociological' conception of class, and in the second, of what may be called a 'logical' or purely formal conception of it. But the ambiguity is not merely terminological; its roots lie deeper, in our traditional social structure, and in our contemporary attitudes to it.

The Backward Classes are a large and mixed category of persons with boundaries that are both unclear and elastic. Together, they comprise roughly one-third of the total population of the country. They are made up of three principal components, the Scheduled Tribes, the Scheduled Castes and the Other Backward Classes. The Scheduled Tribes and Scheduled Castes are well-defined categories, comprising respectively around 7 and 15 per cent of the population. The Other Backward Classes are a residual category; their position is highly ambiguous; and it is impossible to give an exact statement of their number.

The Backward Classes provide a window into modern Indian society as a whole. It has been said about the traditional order of Hindu society that it was so extensively marked by the pre-eminence of the Brahmin, that an understanding of his social situation provided a key to the understanding of its structure as a whole.[5] In many ways the Backward Classes occupy such a privileged position in contemporary Indian society from the point of view of method. For if our interest is in the interplay between equality and hierarchy, there is no significant problem that can escape us if we fix our attention on these sections of Indian society.

If we are to understand how far the law can be used as an instrument of social change, we have to begin by recognizing the disharmony between the legal order with its commitment to complete equality and the social order with its all-pervasive stratification. This disharmony has to be examined in the widest historical and comparative perspective. All modern societies have, in the broadest sense, to contend with the problem of reconciling the ideal of equality with the facts of inequality. It is to this aspect of the modern world that a distinguished European sociologist drew attention when he wrote, 'Modern industrial societies are both egalitarian in aspiration and hierarchical in organization.'[6]

We can better appreciate the paradox of equality in contemporary Indian society by comparing it with other contemporary societies and by contrasting it with societies of the past. We take the ideal of equality so much for granted today that we tend to overlook the point that traditional societies were hierarchical not only in fact but also by design. As Isaiah Berlin had pointed out in a well-known essay on equality, 'Classical thought seems to be deeply and "naturally" inegalitarian '[7]; and both Aristotle and Plato believed in a natural hierarchy of persons, and insisted on appropriate differences of treatment for each of its various levels. Medieval Europe also regarded the hierarchical order to be a part of the natural scheme of things; and, standing on the threshhold of the modern world, de Tocqueville presented a luminous contrast between the 'aristocratic' societies of the past and the 'democratic' societies of the future.[8]

The spirit of hierarchy had its most luxuriant growth in our own traditional society, with its embodiment in the institutions of *varna* and *jati*. Some have been so greatly struck by the stress on hierarchy in the traditional Hindu system of values that they have questioned whether there was in it any appreciation at all of equality as a value. This is to carry a reasonable argument to an unreasonable conclusion, for religious discourse in India never wholly abandoned a concern for man as man, the human spirit behind the external markers of social entitlement and worldly achievement. It is this side of Indian religiosity that attracted Mahatma Gandhi to the message of the Gita and Dr Ambedkar to the teachings of the Buddha.

But it has to be admitted that the idea of equality was for the most part narrowly confined to what may be described as the realm of the spirit. It is difficult from our point of view to see how much solace a man crushed by the burden of an oppressive hierarchy could receive from the thought that his ultimate claim to salvation was as good as that of any other man. At any rate, we have to make a distinction between the spiritual order and the legal order, even though, in a traditional society, both receive their ultimate sanction from religion. In considering the legal order of a society we are less concerned with its ultimate values than with the values that justify and uphold its existing institutions.

The legal order of traditional Hindu society is embodied in the *Dharmashastra*. I am neither a student of law nor a classical scholar, but one does not have to be either to see how radically different their spirit is from the spirit of our Constitution. It can hardly be an accident that the man primarily responsible for its drafting chose as an act of public protest to burn the *Manusmriti* which for two thousand years occupied a pre-eminent position among the *Dharmashastra*.

Shudras and women are marked out in the *Dharmashastra* for indignities of every conceivable kind. They are dealt with more harshly than others; their disabilities are grave and onerous; and they are debarred from most of the ordinary graces of life. The subordination of women is underscored in a well-known verse in the *Manusmriti*: 'In childhood a female must be subject to her father, in youth to her husband, when her lord is dead to her sons; a woman must never be independent.'[9] The Shudra's lot is not much better: 'A Shudra, though emancipated by his master, is not released from servitude; since this is innate in him, who can set him free from it?'[10] Even as sympathetic a reader of the *Dharmashastra* as P. V. Kane was obliged to concede that 'the life of a Shudra was not worth much.'[11]

It would be a mistake to try to account for all this by means of a narrow theory of interests. Some of the injunctions in the *Dharmashastra* are plainly designed to safeguard the interests of the privileged at the expense of the underprivileged. Others seem to express pure and unalloyed malice, as, for instance, the one against the acquisition of wealth by the Shudras on the ground that 'a Shudra who has acquired wealth gives pain to Brahmans.'[12] Above all, one is struck by the luxuriant growth of the discriminatory process which had, in the manner of tropical vegetation, spread in every direction, leaving no ground uncovered.

The Shudras themselves did not remain an undifferentiated category. They became differentiated into superior and inferior, and the discrimination which the Brahmins practised against them was in turn practised by the superior Shudras against the inferior. Shudras came to be dichotomized in several ways, of which Kane mentions three: the dichotomy of *sat* and *asat* Shudras (well-and ill-conducted Shudras); of *bhojyanna* and *abhojyanna* Shudras (those from whom food might or might not be accepted); and of *aniravasita* and *niravasita* Shudras ('clean' and 'unclean' Shudras).[13] In the course of time the last came to be regarded as a separate category, outside the pale of the four *varnas*.

It was for the *niravasita* Shudras—the Chandalas and the Shvapachas—that the worst indignities were reserved. They are the classical forebears of the Scheduled Castes of today. Manu requires that 'the dwellings of Chandalas and Shvapachas shall be outside the village . . . their dress the garments of the dead . . . their food . . . given to them . . . in a broken dish.'[14] Again, what strikes one in all this is not simply that distinctions should be made between superior and inferior, or that they should be made primarily according to birth, but that they should seek to leave no sphere of life free from their impress.

It is a familiar argument among sociologists that no society could possibly function if all the injunctions laid down in the *Dharmashastra* were actually to be practised. More than a hundred years ago, Sir Henry Maine had endorsed the view that the *Manusmriti* 'does not, as a whole, represent a set of rules ever actually administered in Hindostan,' adding that 'It is, in great part, an ideal picture of that which, in the view of the Brahmins, *ought* to be the law.'[15] Custom and common sense obviously played a part in protecting the system from its own absurd conclusions. The fact remains, however, that attitudes towards existing social divisions were radically different then from what they are now; at the same time, many of these social divisions persist, despite the change in spirit encoded in our new legal order. This is the most manifest contradiction in everyday life in contemporary India.

It is easy enough to see the contradiction between the ideal of equality and the practice of inequality. What is far less obvious is that the idea of equality is itself made up of various components which are not always mutually consistent. It might indeed be argued that one reason why there is disharmony between ideal and reality in the modern world is that the concept of equality is itself heterogeneous. I believe that the makers of our Constitution had some awareness of this and of the need to strive for a harmonious construction of the different components of the concept, although there cannot be, in the very nature of the case, any simple formula for achieving such a construction.

The concept of equality is so wide in scope and has had such diverse historical expressions that it would be surprising if it retained a single, univocal meaning. Like all such basic and fundamental concepts, it is both equivocal and inexhaustible. The makers of our Constitution took great pains to incorporate into it whatever they found to be of value in the democratic constitutions of the modern world. They were, in their circumstances, right in doing this, even though breadth of scope had to be achieved at some cost to unity of conception.

Very broadly considered, one can distinguish between equality in the simple sense and equality considered as a ratio. This is an old distinction in western ethical and political philosophy, and Aristotle makes it both in his *Ethics* and in his *Politics*. In his terminology, the distinction is between 'numerical' equality and 'proportional' equality: as he puts it, 'by the first I mean sameness or equality in number or size; by the second equality of ratios.'[16] Aristotle deals primarily with 'proportional' equality, and his whole theory of distributive justice is based on it.

Equality in the simple sense takes no account of the differences among people. It distributes values in such a way that no recipient gets either more or less than any other. 'Every man to count for one and no one to count for more than one': this is the maxim that best sums up the idea of equality in the simple sense.[17] All modern societies try to apply it to the distribution of certain values, and the idea itself is very widely known. In traditional societies too it was acknowledged, but within restricted spheres, as for instance in the domain of kinship and, up to a point, of religion.

But beyond a certain point, a just distribution of values has to take differences between persons into account. Proportional equality consists in maintaining a just proportion while taking into account the differences among persons. If we accept that there are differences among persons and that different persons have to be treated differently, we can still apply the principle of equality, this time not in a simple sense, but by keeping some relevant criterion in mind. Here it is a question of classifying people, and giving every member of each class an equal right to what is allotted to the class as a whole.[18]

But the classification may be made according to merit or according to need, and the implications of the two from the point of view of distribution will be very different. While Aristotle paid much attention to merit, our modern commitment to welfare requires us to pay attention also to need. We know, or believe we know, that people differ according to merit, and if we allot equal rewards for equal merit, we are inclined to feel that the principle of equality has been in some sense satisfied. But people differ also according to need, and we might feel that real, as opposed to formal, equality can be achieved only if we make our distribution proportional to the needs of persons. The first of these two principles may be called the meritarian principle, and the second the compensatory principle or the principle of redress. As we shall see, there are fundamental tensions between the two, although both are invoked in the name of equality.

It would be a mistake to think that these distinctions are made only by professional philosophers or that they are only of academic value. We make them all the time, and, as such, it is important to ensure that we make them consistently and keep their implications clearly in mind. These are the very distinctions we find in Nehru's reflections on equality and inequality set down on the eve of Independence. When he wrote that 'the spirit of the age is in favour of equality' and that 'the spirit of the age will triumph',[19] he had in mind, first of all, the elimination of artificial barriers, such as those of caste, estate or race: this is equality in the simple sense.

But then Nehru went on to say, 'That does not and can not mean that everybody is physically or intellectually or spiritually equal or can be made so. But it does mean equal opportunities for all and no political, economic or social barrier in the way of any individual or group'[20]: or, in other words, reward must bear some proportion to ability, merit or talent. Nor is this all: for 'not only must equal opportunities be given to all, but special opportunities for educational, economic and cultural growth must be given to backward groups so as to enable them to catch up to those who are ahead of them,'[21] which is to say that there must be compensation for need and not just reward for merit.

To say that the spirit of the age is in favour of equality is not to say that there are no critics of egalitarianism, either in this country or in the west. There are critics of egalitarianism among scholars as well as men in public life. The most obvious target of attack is what I have described as equality in the simple sense. It can be made to appear absurd by showing that nobody would seriously wish everyone to be treated equally in all respects. Or, it can be made to appear vacuous by showing that so many qualifications have to be made in order to take differences between persons into account that eventually very little content remains in the maxim, 'Everyone is to count for one, and no one is to count for more than one.'

A few years ago J. R. Lucas, an Oxford philosopher, wrote an article entitled 'Against Equality' in which he attacked the position adopted by the egalitarians, and concluded, 'The central argument for Equality is a muddle.'[22] The argument advanced by Lucas himself is purely formal and pays no attention to the social and historical conditions under which men and women strive to attain equality. More recently a prominent member of the British cabinet, Keith Joseph, has co-authored a book in which he restates the nineteenth-century argument that, since human beings are by nature unequal, it is both futile and perverse to try to establish a social order on the basis of equality.[23]

A strong argument in favour of equality is that equal distribution does not call for any specific justification, whereas any unequal distribution, being but a particular case of unequal distribution, does. Isaiah Berlin puts it thus: 'If I have a cake and there are ten persons among whom I wish to divide it, then if I give exactly one tenth to each, this will not, at any rate automatically, call for justification; whereas if I depart from this principle of equal division I am expected to produce a special reason.'[24] This is not to argue that good reasons can never be found for unequal division. It is only to suggest that what are offered as reasons are not always good reasons but often specious ones, as for example that men should receive more education than women because they have superior intelligence.

Thus, there is a great deal to be said in favour of the idea of simple equality, with all its limitations. The value we place on it can be illustrated by the commitment in our Constitution to the principle of adult suffrage: every citizen, subject to a certain qualification of age, which every citizen is expected to meet in the ordinary course, has an equal right to elect representatives to Parliament and to the State Assemblies (Art. 326). Now that we have this right, we tend to take it for granted, and perhaps also to abuse it. But we have only to turn to our own recent history, or, indeed, to some of our neighbouring countries today to realize that equality in even this simple sense means something, and that people have had to fight in order to achieve it.

If we go back only a couple of hundred years in time we will realize how novel, in historical terms, the very idea of full adult franchise is. The Levellers became famous as an ultra-Republican sect or movement whose leaders supported the people's cause in mid-seventeenth century England. But, as a recent student of that period has pointed out, 'the Levellers consistently excluded from their franchise proposals two substantial categories of men, namely servants or wage earners, and those in receipt of alms or beggars.'[25] And there was no question of extending the franchise to women. Today it requires some effort to see such a movement as a people's movement.

The very idea of citizenship entails an element of equality—equality in the simple sense—that we tend to take for granted because we tend to take citizenship itself for granted. And yet, the right of citizenship—along with 'equality before the law' and 'the equal protection of the laws'—is not something that all Indians have always enjoyed. I know that there are still villages in India where Untouchables do not have the status of full members. If we condemn this violation of our laws regarding citizenship, how much more must we condemn the laws themselves which required that 'the dwellings of Chandalas and Shvapachas shall be outside the village . . . their dress the garments of the dead . . . their food . . . given to them . . . in a broken dish'.

It is not difficult to find the rationale behind the idea of simple equality—citizenship, equality before the law, equal protection of the laws—and we have to consider it, however, briefly. There are undoubtedly differences between persons, and there have to be differences of treatment. From here one might be inclined to argue that for every difference between persons there ought to be a difference of treatment. It is this argument that we must categorically reject as being contrary to both reason and morality. The idea of simple equality merely suggests that differences between persons need not entail differences of treatment; and we can give substance to it only to the extent that we strive to extend those areas of life in which differences between persons are not allowed to interfere with our treatment of them as equal human beings.

To treat people alike, irrespective of outward differences, is to treat them from the human point of view as against the point of view of race, or of caste or of gender. What is meant by treating people from the human point of view can perhaps be best brought home by a consideration of the denial of such treatment. A good example is the attempt to justify the ill-treatment of slaves in the New World by the argument that slaves, being blacks, were not human beings in the full sense of the term.[26] Another example is the omission of the aborigines from the censuses of population conducted in Australia until recently; they simply did not count as human beings.[27] These are extreme examples, but in a country which has been under colonial rule for two centuries it should not be difficult for even the privileged to understand what denial of consideration from the human point of view might signify.

The case for the human point of view in the context of equality has recently been eloquently made by the English philosopher Bernard Williams, and I cannot do better than to refer to his argument here. Williams first draws attention to certain common human capacities, and then goes on to show how important it is to keep these in mind while making an assessment of any kind of social arrangement:

> The assertion that men are alike in the possession of these characteristics is, while indisputable and (it may be) even necessarily true, not trival. For it is certain that there are political and social arrangements that systematically neglect these characteristics in the case of some groups of men, while being fully aware of them in the case of others; that is to say, they treat certain men as though they did not possess these characteristics, and neglect moral claims that arise from these characteristics and which would be admitted to arise from them.[28]

He goes on to add that differences of treatment must rest on some moral principle and not merely on an arbitrary assertion of will.

A student of comparative sociology should hesitate to characterize the arrangements in any society or the reasons offered for such arrangements as arbitrary. At the same time, we cannot but be struck by the nature and number of invidious distinctions recognized and endorsed by the guardians of the traditional legal order in India. No society has allowed such a luxuriant growth of invidious social distinctions as ours. These distinctions of caste, subcaste, sect, subsect and the like have acted over the centuries to smother, if not to efface, the human point of view.

We have to take special care in our consideration of the Backward Classes to keep the human point of view in the forefront. Few groups in history could have suffered from the denial of the human point of view as much or as long as they did. Nor is this all a matter of past history. For two thousand years Untouchables and Tribals have been treated as if they were less than human beings, and this treatment was justified by the argument that they and their children were in their capacities inherently inferior to those born to a superior station in life. These historical disabilities must be kept in mind in any consideration of equality in the new social order.

Principles of proportional equality seek to reconcile the demands of equality with inequalities that are already in existence. As such, they are more complex than the simple equality with which we have been so far concerned. It might be said that in dealing with proportional equality, whether in the context of merit or in the context of need, we are dealing not so much with equality in the strict sense as with justice, or, at best, with equity. But that would be to take too narrow a view of the matter, for the notion of equality must comprehend equality of opportunity and not merely equality of status.

To adopt the human point of view is not to deny or disregard the differences that exist among individuals. If we consider human beings in any real society, we will find that all kinds of differences do exist among them. There are differences that follow from their arrangement in a given social order. There are perhaps also differences that exist independently of this arrangement. For many people the basic and fundamental question is whether there is any correspondence between these two sets of differences.[29]

If we admit that there are differences in capacity and if we accept that there should be differences in reward, then in those matters where these differences are considered relevant, we should ensure a correspondence between the two. I believe that it was Aristotle who argued that to treat equals unequally is unjust, but that to treat unequals equally is also unjust.[30] Although Aristotle was by no means an ardent egalitarian, this particular formulation of the issue would be acceptable as reasonable to most. To take a trivial example, while it would be considered fair to distribute pieces of cake equally among students in a class, it might not be considered fair to distribute marks equally among them.

Every society is characterized by a certain division of labour through which the various activities necessary for collective existence are carried out in an organized manner. From the sociological point of view the division of labour is in some sense what provides each society with its defining features. For example, the division of labour in a traditional Indian village, based on the *jajmani* system, is radically different from what we find in a modern industrial town. The tasks to be performed are different, there are differences in degrees of specialization, and the number of roles and their mutual connections also differ. The division of labour is most clearly manifested in the economic order, particularly in the occupational system, although in a broad sense it may be taken to cover society as a whole.

There are sharp differences of opinion among social theorists about the nature and significance of the division of labour. The French sociologist, Émile Durkheim, took on the whole a positive view of it. Writing in the last decade of the nineteenth century, he noted first and foremost the tremendous expansion of the division of labour since the beginning of the industrial revolution.[31] New occupations had emerged, and old ones had become divided and subdivided into various specialisms. The division of labour was not confined to the industrial field alone, but permeated every area of life, including the arts and the sciences. Durkheim argued that this was a welcome trend, because the division of labour not only brought material progress but also led to increased social cohesion, although he recognized that there were abnormal forms of it which were socially disruptive.

Marx, by contrast, took a very critical view of the division of labour.[32] He saw a close relationship between the division of labour on the one hand, and the capitalist system and commodity production on the other. Marx and Engels regarded the division of labour to be neither desirable nor inevitable. They had a vision of a future society in which no individual would be tied down to a particular occupational role, and each individual would move freely from one occupation to another according to his choice. At the same time, Marx recognized that such a society could not be created directly or immediately out of the existing social order.[33]

The experience of the twentieth century has shown how difficult it is to do away with the division of labour, and I speak not only of the Soviet Union, but also of China. Taking our Constitution and our social structure together, the reasonable position would seem to be that it is not a question of doing away with the division of labour but of regulating it. We can neither attain our economic objectives nor operate our administrative machinery without a properly regulated division of labour. All this requires at the very least a set of rules for recruiting persons to various social positions and for ensuring appropriate rewards for them for the proper performance of their tasks.

To return to Durkheim, his argument is that if the division of labour is to contribute to social well-being, 'it is not sufficient... that each have his task; it is . . . necessary that his task be fitting to him.'[34] However, this objective is achieved only when we have what Durkheim calls a spontaneous division of labour; it is defeated when we have what he calls a forced division of labour. The forced division of labour is, according to Durkheim, a pathological form of it, but one which is likely to be particularly common in societies such as ours.

There is a forced division of labour when external constraints prevent a proper matching of the capacities of individuals with the tasks they are required to perform. We may visualize, on the one hand, a distribution of social positions that together constitute the division of labour; and on the other, a distribution of capacities and talents among the individual members of society. There is perhaps no society which has achieved a perfect concord between the two, but a commitment to such concord is, in my view, a part of our Constitutional commitment to equality. At the same time, our traditional social structure presents a whole series of obstacles to its realization. Family, lineage, clan, caste, sect and gender are what count, rather than individual capacity, in determining which individual will occupy which position in society.

For centuries it has been believed that a man's social capacities were known from the caste or the lineage into which he was born, and that no further test was necessary to determine what these capacities were. And it was considered axiomatic that men and women had radically different capacities not only biologically but also socially. To some extent they did grow up to have different social capacities. But we know today that these differences were a consequence of erroneous beliefs and the artificial social arrangements that rested on them. If we believe that men and women, or Brahmins and Shudras, are born with unequal mental capacities, and if we make unequal provision for their socialization on the basis of this erroneous belief, they will naturally develop unequal capacities as adults.

The scope for equality is severely restricted when women, Shudras, and all kinds of persons are excluded from positions of respect and responsibility in society with no consideration whatsoever for their individual capacities. It is in this context that the principle of equal opportunity acquires its real significance. This principle is written into our Constitution, in its very Preamble as well as among the Fundamental Rights. As I understand it, Art. 16 simply means that no office is too high for any member of society, whether Shudra or Untouchable, so long as there is the ability. Furthermore, equality of opportunity would signify nothing without the concomitant belief that ability or merit or talent might be discovered in any quarter of society whatsoever.

The idea of careers open to talent was a new one that was introduced into Europe in the wake of the French Revolution. It was a revolutionary idea precisely because in the old regime careers were not open to talent, but were determined by birth. The need to throw careers open to talent cannot be too strongly emphasized in our society where status has been more firmly fixed by birth than in any other society at a comparable level of development. The obstacles to free and open competition are many: there are not only objective factors, such as lack of means, but also subjective ones, like lack of motivation, that are a consequence of centuries of organized discrimination.

Today the meritarian principle makes it possible for Untouchables and persons from other disprivileged groups to attain to the highest positions in society, and this is a considerable change from the past. At the same time, it has to be admitted that there is something paradoxical about the principle itself of equality of opportunity. For equality of opportunity demands at best free and open competition. This means that there can be equality only before the competition, but not after it. In other words, equality of opportunity can and does lead to inequality of result, and this must be a source of serious concern in any social order with a fundamental commitment to equality.

It is one thing to recommend equality of opportunity as a way of eliminating discrimination; it is quite another when equality of opportunity becomes only an excuse for relentless competition without any regard for those who lose out in the race. In the west the very people who welcomed the idea of careers open to talent have now begun to be dismayed by the prospects of meritocracy. The publication in 1958 of Michael Young's satirical book, *The Rise of the Meritocracy*[35] struck a responsive chord in many, and some began to wonder whether a meritocratic society can accommodate the human point of view any better than an aristocratic one.

The difficulty of achieving equality of status solely through equal opportunity becomes abundantly clear in our kind of society where the privileged are also overwhelmingly successful in every kind of competition. The fact that the sons of Untouchable labourers generally do much worse in life than the sons of Brahmin civil servants does not prove anything at all about the merits of the parties concerned. What it proves is the difficulty, if not the impossibility, of fully equalizing the external conditions of competition. Indeed, the examination system as well as the job market favours those who start with better resources and better motivation, both of which are products of their superior position in society. Those who lack these need some compensation if there is to be any prospect of achieving substantive as opposed to merely formal equality.

The idea behind the meritarian principle is that society must remove all artificial barriers against free competition, and then leave each individual to find his proper place according to his merit or his deserts. The idea behind the compensatory principle or the principle of redress is that society must intervene in order to ensure that the competition is fair, and not just free. The first principle seeks merely to remove discrimination, and takes little account of the unequal needs of individuals who are unequally placed. The second seeks to take needs into account, and, at the same time, to provide some cushion against the excesses of untempered competition.

The meritarian principle draws its strength from the notion of equality of opportunity, but this notion itself shows a different aspect when we turn from the ideal to the real conditions of competition. In his Rajenda Prasad Memorial Lectures, Justice Mathew had observed, 'In the final analysis, equality of opportunity is not simply a matter of legal equality. Its existence depends, not merely on the absence of disabilities, but on the presence of abilities.'[36] Where these abilities have been damaged or destroyed by the agency of known historical forces, society has an obligation to restore them to life.

It was in the nineteenth century that the meritarian principle came into its own, expressing as it did the spirit of liberal capitalism with its faith in competition and the free market. The compensatory principle has achieved recognition more recently. It owes its inspiration to socialist rather than to liberal thought, and it relies more on the state than on the market for achieving its objective. To the extent that our Constitution has drawn inspiration from both liberal and socialist ideals, both principles are present in it.

What I have designated as the compensatory principle is recognized under different names in different societies. In some East European countries like Poland it is acknowledged in the contrast made between the 'meritocratic' and the 'socialist' principles of remuneration. In India it is known as 'protective discrimination' since it seeks to discriminate in favour of groups that had in the past been discriminated against. It has found a place even in a capitalist society like that of the United States under the rubric of 'affirmative action'. In all cases its objective is to bring about equality under unequal circumstances.

The thrust of the contrast between the two principles of proportionality is well brought out in a survey of educational policy in his own country made by a distinguished Polish sociologist:

> Unequal conditions of cultural life at home cause the unequal cultural development of children from different strata. Hence workers and peasants support the maintenance of a preferential system of access to institutions of higher education for their children, who are less intellectually developed. Given the limited number of educational places, these preferential principles diminish the automatic chances of children of the intelligentsia to enter the spheres of higher education. Consequently, all principles of selection based upon the results of 'purely' meritocratic examinations are especially popular among the intelligentsia. But this in turn automatically reduces the chances of children from worker and peasant families.[37]

We see from the above that a choice between the two principles is not merely a matter of moral judgement; it also involves a conflict of interests.

Both the meritarian and the compensatory principles are to be found among the equality provisions in our Constitution. It seems to me that the primary emphasis in the equality provisions in the part on Fundamental Rights is on the removal of discrimination and the provision of equality of opportunity. The spirit of these provisions seems to be that the individual rather than the caste or the sect is the irreducible unit of society, and that each individual be considered according to his particular merit or capacity or ability. The emphasis in the part on the Directive Principles seems to be somewhat different: here the state is to take into account the special needs of certain strata of society, and to make special provisions for equalizing the unequal conditions obtaining among the different strata.

Perhaps I have overdrawn the contrast between the two parts of the Constitution. Art. 16 itself qualifies the equal opportunity provision by special provisions for backward classes of citizens, and similar provisions have been introduced into Art. 15 by amendment of the Constitution. Nevertheless, there is a difference in spirit between the two types of provision. The first seeks to give free play to merit, the second seeks to accommodate needs. Further, the difference is not simply a difference between merit and need, but between individual merit and the needs of groups or classes of citizens.

If the meritarian principle applies primarily to individuals and the compensatory principle to collectivities, then we have to decide what kinds of collectivities are deserving of special attention. This will depend in part on the structure of groups and classes in the society concerned; but it will depend also on the value assigned in that society to the individual as such. In the United States the blacks, and possibly other ethnic groups, are candidates for affirmative action. In India the groups especially marked out for protective discrimination are the Untouchables and the tribals. In Poland, in the example cited above, the beneficiaries of preferential treatment in the educational system appear to be peasants and workers.

The idea behind protective discrimination and affirmative action is that certain groups, certain castes or races have special claims on society that cannot be sacrificed altogether to the pursuit of individual excellence. At the same time, no society, least of all a modern society in the second half of the twentieth century, can prosper unless it gives an important place to the claims of individual merit. For no matter which community we take and no matter how disprivileged it is, there will be rival claims to whatever it is that is allotted to it to meet its special needs; and it is difficult to see how individual merit can be overlooked altogether in settling rival claims among the individual members of any disprivileged community.

The compensatory principle, as I have labelled it, seeks to articulate a variety of sentiments. It rests on a recognition of existing social disparities as well as their historical basis. Its recognition of existing social disparities is reflected in its concern for the greater needs of some social strata as compared to others. But there is more to it than just this; there is also a sense of making reparation for ancient wrongs, of compensating whole groups and classes for the injuries they have suffered in the past. It is my belief that protective discrimination can and should seek to satisfy present needs; it can do nothing to repair past injuries.

It will not do to exaggerate either the limitations of the meritarian principle or the resources of the compensatory principle by narrowly linking the former with capitalism and the latter with socialism. It is true that historically the strong emphasis on individual merit has been associated with competition, laissez-faire and the free market, in short, with nineteenth-century capitalism; whereas the preoccupation with collective needs, social welfare and the protection of the weak has been associated with twentieth-century socialism. At the same time, no modern society, whether 'capitalist' or 'socialist', can afford to dispense with either principle and rely solely on the other. Even under socialism the educational system and the occupational system must give some place to individual merit as revealed by some form of competition; and even the most aggressively capitalist system has to give the state some role today in protecting the weak by means of compensatory action.

Our hesitation to leave everything to individual merit has grown with our distrust of untempered competition and the market principle. We have grown to value the welfare of the collectivity, and we have learned that some intervention by society is necessary if that is to be achieved. But we must learn also to set limits to this intervention and to see that it does not become merely an euphemism for interference by the state and its bureaucracy. If the weakness of the meritarian principle was that it left too much to the hazards of the market, the weakness of the compensatory principle is that it tends to leave too much to official patronage. It is thus not a question of choosing between the meritarian and the compensatory principles, but of achieving a proper balance between the two.

We come back to the argument that the idea of equality is not a simple or a homogeneous one. There are different components to it, and it is not just a question of reconciling them in the abstract by means of some general formula. No society starts on a clean slate; every society has its own historical legacy. Age-old disparities must be taken into account if the equality in the new social order is to be real and not merely formal. At the same time, we must ensure that in destroying old inequalities we do not create new ones. Discrimination is a dangerous instrument, no matter how pure the intentions are of those who use it. Our own history ought to teach us how infectious the use of discrimination can be, and how careful we have to be in using it even for a desirable end.

These issues came up again and again when what is now Art. 16 was being debated in the Constituent Assembly more than thirty years ago, and I can do not better than to conclude by drawing attention to the observations made by Dr Ambedkar on that occasion. He emphasized the different points of view that needed to be reconciled. There was first the general opinion 'that there shall be equality of opportunity for all citizens'. There was also the view that 'if this principle is to be operative . . . there ought to be no reservation of any sort for any class or community at all.' Then there was the opinion that 'although theoretically it is good to have the principle that there shall be equality of opportunity, there must at the same time be a provision made for the entry of certain communities which have so far been outside the administration.'[38]

What I find most appealing in Dr Ambedkar's own approach to the problem is its reasonableness and its freedom from dogmatism. He insisted on the need to take into account the special claims of certain communities which had for centuries been excluded from positions of respect and responsibility. At the same time, he warned against the possibility that these special claims might 'eat up' the general rule of equality of opportunity altogether. In his own words, 'we have to safeguard two things, namely, the principle of equality of opportunity and at the same time satisfy the demand of communities which have not had so far representation in the State.'[39] It remains to be seen whether discrimination which has in the past been a source of so much evil can, by being inverted, be made a source of good in the future.

NOTES AND REFERENCES

1. P.K. Tripathi, *Some Insights into Fundamental Rights*, University of Bombay, 1972, p. 47.
2. M. R. Balaji *vs* State of Mysore (*A. I. R. 1963 S. C. 649*).
3. P. Rajendran *vs* State of Madras (*A. I. R. 1968 S. C. 1012*).
4. A. Periakaruppan *vs* State of Tamil Nadu (*A. I. R. 1973 S. C. 2310*).
5. See, for instance, L. Dumont, *Homo Hierarchicus: The Caste System and It Implications*, Paladin, 1972. Max Weber also was inclined to argue that the key to the understanding of Hindu society lay in understanding the social situation of the Brahmins; see his *The Religion of India*, The Free Press, 1958.
6. R. Aron, *Progress and Disillusion*, Pall Mall Press, 1968, p. xv.

7. I. Berlin, 'Equality', in his *Concepts and Categories: Philosophical Essays*, Hogarth Press, 1978, p. 99.
8. A. de Tocqueville, *Democracy in America*, Harper and Row, 1966; see also my Kingsley Martin Memorial Lecture, 'Homo hierarchicus, homo equalis' in *Modern Asian Studies*, Vol. 13, No. 4, 1979, pp. 529-48.
9. *The Law of Manu* (trans. G. Buhler), Motilal Banarasidas, 1964, V: 148. Or, again, 'Her father protects her in childhood, her husband protects her in youth, and her sons protect her in old age; a woman is never fit for independence'. Ibid. IX: 3.
10. Ibid., VIII: 414.
11. P.V. Kane, *History of Dharmashastra*, Bhandarkar Oriental Research Institue (Poona), Vol. II, Part I, 1974, p. 163.
12. *The Law of Manu*, X: 129.
13. Kane, *History of Dharmashastra*, Vol. II, Part I, pp. 121-2.
14. *The Law of Manu*, X: 51-2.
15. H. S. Maine, *Ancient Law*, Oxford University Press, 1950, p. 14.
16. Aristotle, *Politics* (trans, Jowett), Clarendon Press, p. 189; the Warrington edition has it thus: 'But equality is of two kinds—numerical and proportionate to desert. Numerical equality implies that one receives exactly the same (i.e. equivalent) number of things or volume of a thing as everyone else. Equality proportionate to desert implies treatment based on equality of ratios.' (Aristotle, *Politics and the Athenian Consitution*, J. M. Dent, 1959, p. 135).
17. See Berlin, 'Equality', p. 81 ff.
18. Ibid. See also Bernard Williams, 'The Idea of Equality' in P. Laslett and W. G. Runciman (eds), *Philosophy, Politics and Society*, Second Series, Basil Blackwell, 1962, pp. 110-31.
19. Jawaharlal Nehru, *The Discovery of India*, Asia Publishing House, 1961, p. 521.
20. Ibid.
21. Ibid.
22. J. R. Lucas, 'Against Equality', *Philosophy,* October 1965, pp. 296-307.
23. K. Joseph and J. Sumption, *Equality*, John Murray, 1979.
24. Berlin, 'Equality', p. 84.
25. C. B. Macpherson, *The Political Theory of Possessive Individualism*, Oxford University Press, 1964, p. 107. Macpherson adds, 'The term servant in seventeenth-century England meant anyone who worked for an employer for wages, whether the wages were by piece-rates or time-rates, and whether hired by the day or week or by the year' (p. 282), and shows that servants and alms-takers constituted a very substantial part of the population.
26. The most comprehensive modern study is G. Myrdal, *An American Dilemma*, Harper, 1944, of which Chapter 4, 'Racial Beliefs', is of special interest here. A classic account is to be found in A. de Tocqueville, *Democracy in America*, Vol. I, Chapter XVIII, 'The Present and Probable Future Condition of the Three Races that Inhabit the Territory of the United States'.
27. F. L. Jones, *The Structure and Growth of Australia's Aboriginal Population*, Australian National University Press, 1970.
28. Williams, 'The Idea of Equality', p. 112.
29. This in a way was the central question posed by Rousseau; see J.-J. Rousseau, 'A Discourse on the Origins of Inequality' in *The Social Contract and Other*

Essays, Dent, 1938. See also my Auguste Comte Memorial Lecture, *The Idea of Natural Inequality*, London School of Economics, 1980.

30. Aristotle, *Ethics*, Penguin, 1978.
31. É. Durkheim, *The Division of Labour in Society*, The Free Press, 1933.
32. See, in particular, *Capital*, Vol. 1; see also K. Marx and F. Engels, *The German Ideology*, Progress Publishers, 1968.
33. K. Marx, *Critique of the Gotha Programme*, Progress Publishers, 1978.
34. Durkheim, *Division of Labour*, p. 375.
35. M. Young; *The Rise of the Meritocracy*, Thames and Hudson, 1958.
36. Reprinted in K. K. Mathew, *Democracy, Equality and Freedom* (ed. Upendra Baxi), Eastern Book Company (Lucknow), 1978, p. 230. Justic Mathew was moved by the tongue as well as the spirit of Tawney, who wrote, 'In reality, of course, except in a sense which is purely formal, equality of opportunity is not simply a matter of legal equality. Its existence depends not merely on the absence of disabilities, but on the presence of abilities.' R. H. Tawney, *Equality* (fourth edition), Unwin Books, 1964, p. 103.
37. W. Wesolowski, *Classes, Strata and Power*, Routledge and Kegan Paul, 1979, p. 133.
38. *Constituent Assembly Debates: Official Report*, Vol. VII, p. 701.
39. Ibid., p. 702.

II

The last lecture was devoted to the argument that there can be different ways of conceiving of equality. Our Constitution itself provides ample support for this argument, requiring us to take merit as well as need into account, and seeking to articulate the human point of view. These ideas, which we now find increasingly difficult to reconcile, were all incorporated precisely because the Constitution was written with a keen awareness of the complex pattern of inequalities actually present in Indian society.

Few will argue that equality in any real sense can be achieved by the sole application of the meritarian principle, without any attention being paid to the unequal needs of persons. Indeed, the limitations of the meritarian principle taken by itself have become apparent to most people in the second half of the twentieth century. In no society are the disparities between the well-and the ill-favoured more conspicuous than in India, and no project for the attainment of equality can hope to succeed if it turns a blind eye to these disparities. It is in this context that the compensatory principle acquires saliency in India; there has to be some discrimination if the weak are to contend with the strong on anything like terms of equality.

The plea for giving all sections of society a fair chance rather than making merit the sole criterion was urged consistently in the Constituent Assembly debates on the equality provisions. One of the Harijan members said: 'The Government can expect necessary qualifications or personality from the Harijans, but not merit. If you take merit alone into account, the Harijans cannot move forward.'[1] Other members spoke about the special needs of the Harijans as a whole, and the various measures that might be adopted to meet them.

Earlier, in dealing with the distinction between merit and need, we came upon another distinction which we then left implicit, but which we must now make explicit. This is the distinction between the individual and the group. The two distinctions—between merit and need, and between the individual and the group—are not the same, but it is difficult to keep them apart in any discussion of equality, particularly in one concerned with protective discrimination. At this stage it is sufficient to say that the meritarian principle tends to take the individual alone into account, ignoring the group or community of which he might be a member. The compensatory principle, on the other hand, tends to identify the individual by the group or community to which he belongs, generally, though not invariably, by being born in it.

The policy of protective discrimination raises two issues which must both be kept in mind in any asessment we make of its successes and failures in India. There is, first of all, the question of how far we are giong to reward merit, and how far make allowance for need. There is also the question, equally important to my mind, of how we are going to balance the claims of the individual with those of the group or the community. We must view with caution a policy of protective discrimination which sets out to decrease the inequalities between castes and communities but ends by increasing the inequalities between individual members of each caste and community. It would be a mistake to believe that by making concessions to castes and communities we *automatically* satisfy the needs of all, or even the most deserving, of their individual members.

The application of the compensatory principle présupposes some kind of classification, and it has been well said in a Supreme Court judgement that 'discrimination is the essence of classification'.[2] The Constitution itself recognizes categories of individuals which it variously describes as the weaker sections or the backward classes. The Scheduled Tribes and the Scheduled Castes have been specified and listed. But there are also other classes or sections which appear to have special claims on the resources of the state either in return for past injuries or on account of present needs. It is impossible to assess these claims without a detailed examination of the structure of Indian society, the various communities, classes and sections of which it is composed, and their mutual relations.

It is easy enough to concede that not merit alone, but need also should be taken into account in the allocation of scarce resources. But there can be competing claims on the same resources, all on the basis of need. Take for instance the claims that might be made to services and posts on grounds of the special needs of individuals and classes: it would be disingenuous to pretend that there is no problem here of balancing the needs of various kinds of persons in meeting such claims. Critics of the meritocracy say, with some justice, that merit is an elusive thing, and that there is an arbitrary element in all judgements of merit. But we must not assume that we all know who has what needs, or that it is always easier to determine the relative needs of persons than it is to determine their relative merits.

The point has to be made at the outset that there are different kinds of needs as well as different ways of meeting them. There are some needs which may be met in such a way that rival claims do not arise, at least not in a direct or obvious way. For instance, one can think of whole sections of society standing in special need of primary education or basic medical care. If these are provided free of charge then the needs of those who cannot afford to pay for them are met without any damage to the claims in these regards of those who can afford to pay.

There are, however, other needs which cannot be met without some judgement being made on the merits of the rival claims. Where the opportunities for employment are few and there are many in need of employment, the claims of some individuals have to be sacrificed in order to meet those of others. This brings out a paradox inherent in the process of discrimination itself. For there is all the difference in the world between a form of protective discrimination from which a disprivileged community, class or section as a whole benefits, and one from which only a few of its individual members benefit. A great deal of what passes for protective discrimination or affirmative action is in fact of the latter rather than the former kind.

If it is true that 'discrimination is the essence of classification' then a great deal will hinge around the classification which determines who the weaker sections—or the backward classes—are. It would be a mistake to assume that such a classification is self-evident, or that it is given to us by the nature of things. It is true that in discussing inequality we often use a geological metaphor and speak of social stratification, as if the whole of society were divided into layers or strata, arranged one on top of another in the way in which the layers of the earth are arranged. But this is only a metaphor which can never do full justice to the complex and fluid patterns in which groups, classes and categories are arranged in a real human society.[3]

Some classifications seem to be organic or 'natural' because they have existed and been acknowledged for a very long time; such, for instance, is the classification of the Hindu population of India into its castes and subcastes. Other classifications appear to be 'rational' rather than natural because they are based on the kinds of impersonal criteria we feel ought to be used for making significant distinctions among persons; such, for instance, is the classification of persons according to their occupation. There obviously is some correspondence between the two, but a problem of choice arises where the correspondence ceases to obtain.

No society has only a single scheme of classification which it uses for every purpose; each has several such schemes among which one or a few may be more extensively used than the others. Broadly speaking, these schemes of classification are of two different kinds. The first uses the individual as its unit, and this gives us classes of individuals according to their income or occupation or education. The second uses the group as its unit, and this gives us an arrangement of clans or castes or other such groups, each having a kind of organic identity of its own. Many have been struck by the subordination of the individual to the group in our own traditional society. Nehru, for instance, described its structure thus: 'This structure was based on three concepts: the autonomous village community, caste, and the joint family system. In all these three it is the group that counts; the individual has a secondary place.'[4] These groupings, among which we shall be concerned primarily with castes, maintained an identity over and above that of their individual members, and perpetuated themselves by a kind of universal succession.

The distinction between societies that assign priority to the group in their classifications and those that assign priority to the individual corresponds in large measure to the distinction made famous by Henry Maine between societies based on status and those based on contract.[5] We have inherited a social order based on status, one in which the individual did not count for very much. 'Equality of opportunity' will mean very little if at every turn the individual is shadowed by the caste or the community into which he was born. Justice must be rendered to the castes and communities which have in the past been denied justice; but if we do this without any regard at all for the cost to the individual, instead of moving forward into the new social order promised by the makers of our Constitution, we might move backward into the Middle Ages.

Any discussion of the structure of Indian society must begin with a consideration of the inequalities that are to be encountered in almost every sphere of life. India has been viewed as a text-book example of a hierarchical society. If we take traditional institutions such as caste, village community and joint family, we will find that each is constituted according to a hierarchical design. The new economic forces have not fully effaced this design, but have on the other hand added other inequalities to those already in existence.

The most notable feature of inequality in Indian society from the past to the present is its visibility. Even though inequalities exist in all complex societies, they are in general more visible in agrarian as compared to industrial societies. Such inequalities are visible in the settlement pattern of our villages where the poor and the ill-favoured live apart from the rich and the well-born. Even in our large cities it is impossible not to be struck by the physical distinctions among persons in their dress, appearance and deportment. Some of these distinctions are a consequence of widespread poverty; in a relatively affluent society there is a more even distribution of the basic amenities of life. But there are other distinctions that derive from the peculiarities of our traditional social order.

Social distinctions are not only more visible in India, they tend to be, on the whole, more rigid. By and large, individuals live and die in the station of life into which they are born; marriage also is fairly strictly regulated. There is little mobility, and the barriers between the classes and strata appear to be almost insurmountable. In a society in which the individual moves more or less freely from one level to another in the course of his life, his individual identity appears more important than the class or stratum to which he might belong at a particular moment. In a society characterized by immobility, on the other hand, an individual's personal qualities appear to be of less account than the group of which he is a part. This subordination of the individual to the group is a feature of our traditional social order to which we have already drawn attention; it is a feature that does not harmonize very easily without new legal order.

Inequalities are not only visible and rigid, they are also highly elaborate. The social distance between the top and the bottom of the hierarchy is very large, and there are numerous grades in between. When we look at our traditional caste structure, we are struck by the divisions and subdivisions within it. When we look at our traditional agrarian hierarchy, we are struck equally by the number of intermediaries that stood between the landlord and the tiller of the land. The proliferation of invidious distinctions is a feature also of our modern social life. It has been said that many of these new distinctions are an artefact of colonial rule; even so, colonial rule found in our society a particularly fertile soil for generating distinctions of rank.

How are we to account for these inequalities whose existence is acknowledged by everyone? When and how did they originate, and what has sustained them over the centuries? Opinion is sharply divided on the question of the key to the problem of Indian inequality, and one may well ask whether there is in fact a single key to it. I shall consider two alternative approaches to the problem, because they indicate two different ways of identifying those who are most in need of special assistance.

For some the key to the problem of social inequality lies in the domain of material factors. They would say that there is pervasive inequality in India because of its all-round poverty, its general economic backwardness, and its slow rate of economic growth. Extremes of wealth and poverty are characteristic features of economically backward societies. A high rate of economic growth, on the other hand, creates the kinds of opportunities through which the barriers between classes and between strata become dissolved. A stagnant agrarian economy, long under colonial domination, has had very little scope for the loosening of its social rigidities.

The argument about the linkage between economic backwardness and social inequality has been made in a broad comparative and historical perspective by Gunnar Myrdal.[6] Myrdal contrasts the economically-backward societies of Asia with the economically-advanced societies of the west, and finds that, while inequalities exist in both, they are more visible, more rigid and more elaborate in the former than in the latter. Whether we attribute India's economic backwardness to its traditional institutional structure or to its prolonged subjection to colonial rule, there is no denying the fact that this backwardness has contributed much to the creation and maintenance of a very rigid system of social inequality.

The same contrasts are revealed when we examine the relationship between economic development and social inequality in a historical perspective. Western societies were not always characterized by high rates of economic growth. Pre-industrial society in Europe was, compared to the present, economically backward, and with this backwardness was associated the existence of all kinds of invidious social distinctions, although it is perhaps true that neither the degree of economic backwardness nor the extent of social inequality there was ever as great as in India. With the creation of new economic opportunities in the west, many of the traditional social distinctions began to dissolve, and a fairly fluid system of classes came to replace the rather rigid system of estates characteristic of the past.

Those who assign primacy to the economic dimension of social inequality tend to view the problem of backwardness in India as being different only in degree from similar problems elsewhere. Perhaps the extent of poverty is greater in this country than in other countries; perhaps more people, in both relative and absolute terms, are in need of special economic assistance here than elsewhere. From the viewpoint of the planner and the policy maker there would appear to be certain advantages in defining backwardness in purely economic terms; one can then measure its extent, and apply uniform rules for deciding how assistance may be matched with need in every individual instance.

But not everybody regards the problem of social inequality in India to be basically an economic problem. There are those who maintain that if we are to get to the root of inequality in India we must begin with what is unique to Indian society and not with what it has in common with other societies. They point out that disparities of wealth and income exist in all complex societies, whereas untouchability exists only in India, and that unless we understand the social basis of untouchability we will never be able to find a solution to the problem of backwardness in India. In this view poverty, and even destitution, is only an aspect of a larger problem which has its roots in the very structure of traditional Hindu society.[7]

Nobody can deny the special significance of the caste system for the problem of inequality in India, including contemporary India, From the sociological point of view, the caste system has a morphological aspect and an ideological aspect, and to a large extent the one has reinforced the other. Morphologically, the whole of Hindu society has been divided and subdivided into a large number of small and well-defined groups, ranked in an elaborate and complex hierarchy; it was these groups, rather than individuals, which constituted the building blocks of caste society. Ideologically, there has been a strong emphasis on collective as opposed to individual identity, and on hierarchy, particularly as viewed in terms of the opposition between purity and pollution; social superiority was defined not so much in terms of wealth as of purity, and the stigma of pollution rather than poverty was what defined social inferiority.

Indian society may be represented in terms of *either* its class structure *or* its caste structure. Each representation provides or at least claims to provide a kind of global perspective on Indian society. Those who advocate the perspective of class seek to explain by it not only inequality in its various forms but every other important aspect of life as well, from politics to religion. Alternatively, those who favour the perspective of caste argue that caste permeates every sphere of collective life in both its morphological and ideological aspects.

Sociologists who believe that caste provides the ground plan of Indian society maintain that 'class' is a category of capitalist society, or industrial society, or western society, and that Indians themselves do not perceive their social world as being divided into classes as Europeans or Americans might do; one variant of this argument is that in India 'class' is nothing but a particular grouping of castes. The argument on the other side is that the perspective of caste is a backward-looking perspective on Indian society, that caste might have been significant in the past, but that today it is merely a shell that conceals the real cleavages in Indian society, which are those of class.

I believe the question to be of such fundamental importance to the application of what I have called the compensatory principle, that I would like to devote some attention to it, however briefly. There are two obstacles to a clear understanding of the distinction between caste and class, Firstly, there is no general agreement as to what people mean by class, and, to some extent, even caste. Secondly, there is considerable correlation at the empirical level between caste and class, which inclines peoples to the somewhat misleading conclusion that caste is an aspect of class, or vice-versa.

The distinction between caste and class, it appears to me, is drawn differently in the legal as against the sociological literature. The main reason behind this is that the lawyer tends to think of 'class' in a rather different way from the social scientist. Being a sociologist, I will not presume to make a judgement on these divergent conceptions of class. At the same time, important decisions of the Supreme Court refer to the sociological factors that are to be taken into account in defining the Backward Classes, and this encourages me to confront the legal with the sociological conception of class.

From the sociological point of view the legal conception of class appears to be very broad and very general. For the lawyer class is a 'logical' rather than a 'sociological' category. In this sense a class is a category that we get by any kind of logically consistent classification. Clearly there is a sense in which we can talk about classes of numbers in mathematics, or of classes of phonemes in linguistics. The lawyer speaks of 'classes' of persons, and he is satisfied so long as the classification is reasonable in terms of the objective behind a particular Constitutional principle or legislative enactment. From his point of view it makes sense to describe as classes not only landowners, tenants and labourers, or upper-, middle- and lower income groups, but also Tribals, Untouchables and other groups of castes and communities defined in a particular way for a particular object.

The sociologist tries to give a more restricted meaning to the concept of class.[8] He does not regard landowner, tenant and labourer on the one hand, and Brahmin, Jat and Chamar on the other, as being categories of the same kind. For him only the former constitute classes in the true sense of the term, and not the latter. The fact that most Chamars are agricultural labourers or that most Jats are cultivators does not make them into classes, for the identity of the first set of catgories has a different basis from that of the second.

It is far from my intention to suggest that sociologists themselves are in complete agreement on the meaning of class. There is, first of all, the well-known difference of approach between Marxist and non-Marxist sociologists.[9] The Marxists tend not only to assign overwhelming significance to class, but also to define it in a particular way. Non-Marxists are, on the whole, more eclectic in the choice of criteria for defining class. I think it would be fair to say that sociologists are more in agreement on what should not be reckoned as class than on what class actually is.

There are two interrelated components to what may be considered as the common core of the sociological conception of class. The first relates to the kinds of criteria by which classes are differentiated from each other, and the second to the kinds of units with which classes are constructed by the application of such criteria. The criteria used for differentiating classes are economic criteria, and classes are made up of individuals who have only their economic conditions in common. The importance of the economic criterion in the definition of classes will be readily acknowledged. The significance of starting with the individual in reckoning classes is less easily recognized; it becomes manifest only when we contrast a system of classes with a system of castes.

It is not enough to say that social classes should be defined by economic criteria, because one can think of several economic criteria which are not all of the same kind. Here again the Marxist viewpoint is distinctive because it insists that, objectively, classes should be defined in terms of the sole criterion of the ownership or non-ownership of the means of production. But others have pointed to the independent importance of occupation and income.[10] Among those who own no land or capital there may be some who are in superior occupations and earn high incomes; conversely, some property owners may have smaller incomes and lower prestige than some individuals in high-salaried occupations who may not own any property at all. At any rate, nobody can deny that in the contemporary world an individual's income and occupation—and perhaps also his education—are good indicators of his needs, and of his capacity to meet the needs of the members of his household.

Whichever way we look at it, a class is an aggregate of individuals (or, at best, of households), and, as such, quite different from a caste which is an enduring group. The distinction between an aggregate of individuals and an enduring group is of fundamental significance to the sociologist, and, I suspect, to the jurist as well.[11] A class derives the character it has by virtue of the characteristics of its individual members. In the case of caste, on the other hand, it is the group that stamps the individual with its own characteristics. There are some affiliations which an individual may change, including that of his class; he cannot change his caste. At least in principle a caste remains the same caste even when a majority of its individual members change their occupation, or their income, or even their relation to the means of production; it would be absurd from the sociological point of view to think of a class in this way. A caste is a grouping *sui generis*, very different from a class, particularly when we define class in terms of income or occupation.

The irreducible identity of castes in Indian society is acknowledged by sociologists as well as lawyers. I interpret Justice Hegde's statement in a Supreme Court judgement that 'A caste has always been recognized as a class'[12] to mean simply that the existence of castes must be acknowledged as a significant part of our social reality. Yet there is a certain uneasiness about this acknowledgement in view of our commitment to a casteless society. I see a trace of this uneasiness in the statement in another recent Supreme Court judgement that the Scheduled Castes 'are not a caste within the ordinary meaning of caste.'[13] It is as if we were forced to acknowledge the existence of castes, wishing at the same time that they were classes!

A sociologist unfamiliar with the intricacies of Indian social structure is likely to detect a certain anomaly between the title of Part XVI of the Constitution and the provisions actually made under its various articles. For, while the title speaks of 'Special Provisions Relating to Certain Classes', all the articles except one deal with the Scheduled Castes, the Scheduled Tribes and the Anglo-Indian community. Where references are made specifically to the Backward Classes, as in Art. 338 (3) or in Art. 340, it is not altogether clear that they are conceived of as being different in kind from the Scheduled Castes, the Scheduled Tribes and the Anglo-Indian community. Again, it is perhaps a characteristic of our predicament that, in our Constitution as well as in everyday life, when we say 'class', as often as not, we mean 'community'. Even where the spirit of the age wants us to attend to the individual, castes and communities are forced on our attention by our traditional social structure.

Some students of the Indian Constitution have argued that the only reasonable classification in the Indian context is the classification into castes and communities. This point of view has been put forward most forcefully by L. G. Havanur as Chairman of the Karnataka Backward Classes Commission. He states in his Report that:

> Class is synonymous with caste or tribe, so far as Hindus are concerned.
>
> Class is synonymous with tribe, or racial group, so far as tribal communities are concerned.
>
> Class is synonymous with section or group so far as Muslim, Christian and other religious communities and denominations are concerned.[14]

This would seem to leave very little room in the new legal order for any mode of classification other than those inherited from the past.

Much as one may disapprove of the categorical manner in which Mr Havanur presents his case, one cannot ignore the wealth of legal, historical and sociological material he has presented in support of it. This material shows the salience of castes and communities not only in our ancient past but also in our more recent history. It is true that the traditional legal order emphasized the group at the expense of the individual; but there can be little doubt that this emphasis was given an additional edge by the manner in which the British transferred power to Indians after two centuries of colonial domination.

Mr Havanur has shown how from the very beginning the leaders of the Indian National Congress sought to articulate the demands of the various classes in Indian society, by which they meant Muslims, Christians, Sikhs, Parsis, Brahmins, Depressed Classes, etc. The term 'Backward Classes' began to acquire currency from around 1919, and to be used broadly to include the Depressed Classes, the Aboriginal Tribes and the Other Backward Classes. 'The word *class*, besides being used in official and semi-official documents, was also being used by political leaders, social reformers and the like to apply to castes, tribes and communal groups.'[15] In the case of Muslims, Christians and Parsis there could be very little scope for confusing the loose meaning of 'class' with its strict socio-economic definition; in the case of the Backward Classes some scope for such confusion obviously exists.

In the three decades preceding the formation of the Constituent Assembly the divisions into castes and communities that had existed in Indian society from ancient times acquired a new kind of legitimacy through the political process. Communal politics, minority politics and the politics of backwardness became closely intermeshed. It is in this light that we have to interpret Mr Havanur's laconic statement, 'Caste has come to stay'.[16] The British made their contribution to the crystallization of the political identities of castes and communities in the emerging social order. This was partly because they felt that they had a special responsibility in protecting the interests of the minorities and the Backward Classes in the competition for power. But it was also because the demand for self-government could be kept at bay by playing one community off against another.

The balance of power between castes and communities was an important concern for those who participated in the deliberations of the Constituent Assembly. The equality provisions that came to be written into the Constitution, particularly the provisions relating to protective discrimination, cannot be understood in isolation from this concern. Those to whom special provisions were to apply were at first conceived in a broad way, to include a variety of castes and communities; classes in the strict sense of the term hardly figured in this consideration. There were those who wanted the religious minorities to be included, and there were those who argued for the inclusion of virtually all Non-Brahmin Hindu castes. When Dr Ambedkar proposed that benefits be reserved for the backward classes, it was argued against him that if this was done and if the backward classes were defined in a limited and restricted manner, the special claims of millions of others would be overlooked.[17]

The special claims of the Scheduled Castes and the Scheduled Tribes arise out of the conditions under which they have been constrained to live from ancient times. The defining feature of their condition was that they were in many important regards placed *outside* the bounds of the larger society, the Scheduled Tribes on account of their isolation in particular ecological niches, and the Scheduled Castes on account of the segregation imposed on them by the rules of pollution.[18] There were in the past, as there are at present, many different tribes among the Adivasis and many different castes among the Harijans, but they all shared in common the condition of being in one sense or another exterior to the larger society.

The exteriority of these two groups of communities puts their claims on a totally different level from the claims of all other communities in Indian society. Most of the disprivileges from which they have suffered and many of those from which they still suffer can in one way or another be related to it. So long as this condition exists the very possibility of creating equality in the external conditions of competition is denied.

It has to be emphasized that the disabilities from which the Harijans and the Adivasis suffered were in each case imposed on the community as a whole, and not on individual members of particular communities. This was notably so in regard to the stigma of pollution which was the lot of every Harijan caste in traditional Hindu society. P. V. Kane has brought out well the distinction between individual segregation and collective segregation on account of pollution.[19] The segregation of the individual, of no matter what caste, on account of the pollution of birth or death was temporary; quite different was the segregation of whole communities on account of the pollution that was imposed on them as a permanent and inescapable condition of life.

The isolation of the tribal communities was likewise a collective and not an individual affair. Its isolation enabled each tribal community to retain its own social organization, its own customs, its own religion and, above all, its own language. It also imposed on its members a rather low level of material existence. Indeed, what came to be identified as the tribal population of India in the nineteenth and twentieth centuries consisted precisely of those communities which by virtue of their material and cultural isolation had remained outside the mainstream of national life. Their collective deprivation has been in every way as marked as that of the Untouchables.

Although the deprivations traditionally suffered by the Harijans and the Adivasis were different in their specific manifestations, there is a certain logic in treating them together from the viewpoint of protective discrimination. Those who have been kept out require special facilities to be brought in. Special care has to be taken to ensure that they are able to exercise their rights as full citizens in the new legal order. Above all, no cost should be counted too high if it ensures the widest diffusion of literacy and education among them. Since disabilities have been imposed on entire communities, those measure should have the highest priority which directly benefit the largest number of individual members of these communities.

In a very broad sense each caste fixed its collective identity on its individual members and all castes had a position inferior to the Brahmins. Some such argument has been made to claim special concessions for a whole range of castes and communities which occupied the middle levels of the traditional ritual hierarchy. The implication of this is that our first priority ought to be to bring about equality between castes before we attempt to bring about equality between individuals. This, it seems to me, is the spirit behind Mr Havanur's somewhat unusual interpretation of the Constitution: 'Hence the Constitution suggests *recognition of castes for their equalisation.*'[20]

It would be wrong to argue that in providing for equal opportunity, or even the equalization of opportunities among those unequally placed, we should always give priority to the group over the individual. Such a course might be justified if all the groups with which we are concerned—castes and communities—were absolutely homogeneous on every significant scale of inequality. But we know perfectly well that they are not. Firstly, a caste which has a low ritual status may be materially well off; secondly, individual members of most castes vary considerably in their actual material condition, irrespective of the traditional ritual status of their caste, It is only at the very lowest end of 'he scale that the assumption of uniform deprivation holds true to a large extent.

The assumption of perfect congruence between the collective ritual status of a caste and the actual material condition of its individual members does not hold good today, and probably never held good in the majority of cases even in the past. Various forces are at work today which increase the dissociation between caste and income, caste and occupation, and caste and education. These forces draw the individual relentlessly away from the power and the protection of his caste. They compel us to take more and more account of the needs of the individual irrespective of his caste, for his caste tells us less, and less about the total range of his deprivations. The new legal order must make provision for the individual to bring his needs to the attention of the state in his own right, without the mediation of his caste.

To the understanding of a sociologist, the very concept of citizenship in the Constitution of India is that it is an unmediated relationship between the individual and the state. This is a modern concept, characteristic of societies of a particular kind, and not a universal feature of all human societies. The modern concept of unmediated citizenship may be contrasted with the pattern prevalent in traditional West African societies where, Meyer Fortes tells us, 'it is a fundamental prnciple of Ashanti law that lineage membership is an inextinguishable jural capacity and the basic credential for citizenship.'[21] It was much the same in traditional Hindu society: the individual was a member of society by virtue of his membership of a caste which he acquired by birth into a particular family.

Now it is one thing to make provisions of a specific nature and for a limited duration for the Scheduled Castes and Scheduled Tribes in order to protect them from injury and abuse, and in order to ensure that the conditions under which they compete with other members of society are fair and not just free. It is quite another thing to make the scope of protective discrimination so extensive that in every case, or in almost every case, the caste to which an individual belongs becomes a relevant factor in determining his entitlements. For the idea of citizenship as inscribed in our Constitution is the very antithesis of the traditional idea that caste membership 'is an inextinguishable jural capacity and the basic credential for citizenship.'

The meritarian principle, as I noted earlier, has been closely, and, no doubt rightly, associated with individualism, and we cannot ignore the many sins that have been laid at the door of individualism, especially by the advocates of socialism. But we cannot throw out individualism root and branch merely on account of its excesses or its perverse expressions. We may not share all of Durkheim's enthusiasm for individualism as the source of a new religion and a new morality,[22] but we must not hesitate to acknowledge what we owe to the individual. Above all, in the context of Indian society, here and now, we must realize that the alternative to individualism may not be the cherished dream of socialism, but a moral order in which the individual is once again displaced by clan, caste and community.

The shadow of the community loomed large over the minds of many of those who had assembled to prepare the Constitution. In particular, there was much sentimental attachment to the idea of the traditional village community. While presenting the Draft Constitution, Dr Ambedkar attacked the village community and said, 'I am glad that the Draft Constitution has discarded the village and adopted the individual as its unit'.[23] In the debate that followed many were vocal in the cause of the village, but few spoke up for the individual; and yet what they wrote down would signify little without the individual being given a place in the centre.

Those who view inequality in terms of the hierarchy of castes tend to emphasize ritual status, because it was this rather than income or wealth or even occupation that was fixed at the same level for all individual members of a caste. On the other hand, there are those who view inequality in economic terms, and they tend to emphasize the distribution of income and wealth among individuals. We now have a fairly large number of studies of poverty and income distribution which tend to present a somewhat different picture of inequality in Indian society from the one we have just been considering.[24] Unfortunately, there has been very little synthesis of the work done on inequality by sociologists whose emphasis is on the hierarchy of castes, and the work done by economists whose emphasis is on the distribution of income between individuals and between households.

However strongly one might feel about the rigidities of caste, it will generally be conceded that these are no longer as severe as they were even a generation ago. The stigma of pollution, the segregation of Untouchables and the isolation of Tribals persist in practice, no matter what the law lays down, but even the practice is less uniform and less rigid than in the past. The sense of distance between castes considered superior and inferior has in general become attenuated: at any rate, no sociologist will seriously argue that the social distance between castes that was a part of the traditional order is now on the increase. It is otherwise with economic inequality and poverty. The absolute number of those below the poverty line is increasing, and there is no clear evidence that their proportion in the population as a whole is declining.

Whereas sociologists talk about the social backwardness of castes and communities, economists emphasize the material poverty of classes of individuals. The depth and extent of poverty, and the size of the classes in its grip have received expert attention from economists only recently. There are enormous difficulties in arriving at agreed measures of poverty in a country such as ours. Statistics on income and expenditure are difficult to collect in a predominantly agrarian economy, and, where levels of living are so low, there is room for disagreement among experts on where the line of poverty in an absolute sense ought to be drawn. There is, however, general agreement that substantial sections of our population—anything between 25 and 40 per cent of the total—live under conditions of extreme poverty.

Poverty no doubt existed in this country in the past as well, and some of its roots clearly go back to the traditional social order. But there are aspects of it that are equally clearly of more recent origin. It is doubtful if India ever witnessed such a massive concentration of poverty as one may see today in cities such as Calcutta, Bombay and Madras. Properly speaking, these are neither industrial nor preindustrial cities; they owe their origin to the economic demands of capitalist expansion under colonial dominance. It may be argued that concentrated poverty has increased not only in the urban but also in the rural areas, and similar forces have been responsible for their increase in both cases.

The forces that lead to the concentration of poverty and the increase of economic inequality between individuals are also the ones that disrupt the traditional structure of castes and communities. Capitalism creates new inequalities, but it also undermines old ones. The economic forces that push people below the line of poverty do not pay much regard to the finer points of the traditional distinctions of status. Those who move into the slums of a large city leave behind a part of their traditional identity, including their traditional concerns for purity and pollution. The pavement dwellers of Calcutta include persons from all castes and communities, though not in equal proportions; it would be unrealistic to believe that old distinctions can survive unchanged under these new conditions of life.

It has often been said that, despite its iniquities, the traditional order assured a measure of economic security to members of all castes, including the lowest. Each caste was assigned a specific occupation, and its members had a kind of hereditary right over that occupation. Economic relations were governed by status rather than contract, and the community as a whole had an obligation to see that all its members had some gainful employment. This is no longer the case either in principle or in practice. Employment is determined by the impersonal laws of the market which every year add to the number of unemployed individuals from every caste and community.[25] One's life chances, including one's employment prospects, are no longer guaranteed by caste, although they may be greatly improved by birth in a wealthy family of no matter what caste.

Traditionally the status of a caste was closely linked to the occupation over which its members had hereditary rights, but the nature of the relationship between caste and occupation has been much misrepresented. A caste occupation, properly speaking, is the occupation traditionally associated with the caste as a whole, and not the occupation actually practised by its individual members. It is doubtful that there was at any time a complete correspondence between the two. At any rate, even before independence many castes, and probably most, had more than half their working members in occupations other than those specifically associated with their caste.[26]

It should be clear that even when one acknowledges the desirability of limiting the claims of merit by those of need, it is not easy to determine what kinds of needs there are and how they are to be met. It is one thing to try to satisfy needs through the provision of general facilities such as those relating to nutrition, health and literacy, and quite another to make special provisions, such as those relating to job reservation, that can satisfy directly only a few individual members of groups that are made up of very many. In particular it is fallacious to argue that the equalization of castes can be achieved by means of job reservations in the government. Such jobs are too few in number to materially alter the conditions of any caste as a whole; and there is little reason to believe that the personal advancement of an individual and the social betterment of the caste into which he was born have very much to do with each other.

In matters such as job reservation we have to consider seriously how much weightage to give to the material condition of the individual candidate and how much, if any, to the social status of the caste to which he belongs. Attempts have been made, as in the state of Kerala, to apply a means test in addition to the test of caste or community, but the result does not appear to have been materially different from what one might get by applying the caste test alone.[27] It is now becoming increasingly clear that in seeking to do justice to castes or communities we might deny justice to individuals, for we cannot any longer pretend that all or most or even many of the needs of the individual will automatically be taken care of by his caste, once the status of that caste is enhanced. The ends of justice are hardly met if our vain endeavour to bring about equality between castes leads only to the increase of inequality among the individual members of every caste.

Leaving aside the very special claims of the Scheduled Castes and the Scheduled Tribes, we have to concede that poverty is a very serious problem in Indian society, and that in choosing its victims it does not necessarily discriminate among castes. Justice Gajendragadkar's statement that social backwardness is 'the result of poverty to a very large extent'[28] is perhaps even more true today than when it was made nearly twenty years ago. More recently Jagjivan Ram has observed that 'Problems of a poor Brahmin and a poor Harijan are the same'.[29] Certainly the problems of a poor Brahmin may be more acute than those of a well-to-do member of a 'backward' caste demanding a place in the administration on the plea that his community has fewer than the average number of members in it.

To argue that a 'poor Brahmin and a poor Harijan'; should be treated alike is to assert what I have earlier described as the human point of view. It is to maintain that considerations of race, caste and creed should be set aside when we are faced with the real needs of individual human beings. It is useful to remember that the human point of view may be asserted also on behalf of a Brahmin; and it is salutary to be reminded of this by a Harijan leader of national importance. In a caste-ridden society nothing is easier than to assert the human point of view on behalf of the members of one's own caste.

Clearly, in the case of the Untouchables, backwardness is not solely a matter of poverty, the extent of which might vary from one individual to another. It is due also to the stigma of pollution which attaches to the caste or community as a whole. Attitudes to pollution die hard in our society, and they reappear in the form of social prejudices which certainly weigh against the Untouchables, often heavily, in most competitive situations. The position of other castes which claim job reservation on a par with the Untouchables is hardly comparable. They cannot collectively claim to be victims of prejudice on account of their traditional status as Harijans and Adivasis can. In their case the particular circumstances of the individual claimants, rather than the status of the caste as a whole, must be considered decisive.

The prospects of material advancement through job reservation have led to kind of competition for backwardness among castes at the middle levels of the hierarchy. This kind of competition creates a vested interest in backwardness, and it combines the worst features of a hierarchical and a free-market society. It stifles individual initiative without creating equality between individuals, and it obstructs the natural processes through which the barriers between castes and communities can be effaced. By making caste and community a relevant factor in every sphere of activity, it pushes the human point of view into the background.

The ideal of equality has at best a very insecure foothold in our society. It can never become securely established until we reject the distinctions of caste in all their implications. It is a mistake to try to separate the morphology of caste as a set of self-perpetuating groups from its ideology. The morphology and the ideology of caste are closely linked, and the ideology is totally antithetical to the values of equality. Thus any effort to strengthen the identity of castes with a view to their equalization cannot but end in failure.

It is not true that the hierarchical values on which the distinctions of caste rest have never been challenged in Indian society before the modern age. It has often been pointed out that the first great protagonist of equality, the Buddha, was himself born in this land of hierarchy. Throughout the Middle Ages there was a succession of religious reform movements which challenged the established hierarchy of caste in the name of equality among men. But they all came to grief because of their inability to cut through the existing divisions of society. Describing the course of the movement for equality started by Chaitanyadeva in eastern India, N. K. Bose wrote, 'The ideas propounded by Lord Chaitanya remained confined to particular sects; they were not able to break down the intolerance embedded in society as a whole and usher in a new flood of life. The Vaishnavas were in effect transformed into a new caste.'[30] This seems to have been the normal cycle of development: what started as a movement for social reform became hardened into a sect which became transformed into a caste which then found a place for itself in the established order of castes.

India has been described as the land of 'the most inviolable organization by birth',[31] and the subordination of the individual to the group is an inseparable aspect of this organization. It is here more than anywhere else that we have to be vigilant about claims made by individuals on the strength of their birth in a particular group. This society made a terrible mistake in the past in believing that merit was an attribute not of individuals but of groups, that being born a Brahmin was in itself a mark of merit. We shall make the same kind of mistake if we act on the belief that need too is always, and not just in special cases, an attribute of groups rather than of individuals.

NOTES AND REFERENCES

1. *Constituent Assembly Debates: Official Report*, Vol. VII, p. 688.
2. Ray CJ in State of Kerala *vs* N. M. Thomas (*A.I.R. 1976 S.C.* 497).
3. For a discussion of the distortions of the geological metaphor see A. Béteille, *Inequality among Men*, Basil Blackwell, 1977, Chapter 6.
4. J. Nehru, *The Discovery of India*, Asia Publishing House, 1961, pp. 247–8. See also N. K. Bose, *The Structure of Hindu Society*, Orient Longman, 1975; and L. Dumont, *Homo Hierarchicus: The Caste System and Its Implications*, Paladin, 1972.
5. H. S. Maine, *Ancient Law*, Oxford University Press, 1950.
6. G. Myrdal, 'Chairman's Introduction' in A. de Reuck and J. Knight (eds.), *Caste and Race: Comparative Approaches*, J. & A. Churchill, 1967, pp. 1–4; see also his *Asian Drama: An Inquiry into the Poverty of Nations*, Penguin, 1968.
7. The sharpest recent statement of this point of view is to be found in Dumont, *Home Hierarchicus*.
8. Sometimes the social anthropologist also uses the wider conception of class, as, for instance, in talking about marriage 'classes' among the Australian aborigines. But this usage is today acknowledged to be archaic, if not eccentric. See C. Lévi-Strauss, *The Elementary Structures of Kinship*, Alden Press, 1969.

9. For a discussion of various conceptions of class (and of 'classlessness'), see S. Ossowski, *Class Structure in the Social Consciousness*, Routledge and Kegan Paul, 1963.
10. See, for instance, G. D. H. Cole, *Studies in Class Structure*, Routledge and Kegan Paul, 1955.
11. Some sociologists draw their support for the distinction from the legal literature. See, for instance, M. Fortes, *Kinship and the Social Order*, Routledge and Kegan Paul, 1970, Chapter XIV.
12. Hedge, J. in A. Periakaruppan *vs* State of Tamil Nadu (*A.I.R. 1973 S.C.* 2310).
13. Ray CJ in State of Kerala *vs* N. M. Thomas (*A.I.R. 1976 S.C.* 501).
14. Karnataka Backward Classes Commission (Chairman, L. G. Havanur), *Report*, Government of Karnataka, 1975, Vol. 1, Part 1, pp. 98–9.
15. Ibid., p. 58.
16. Ibid., p.iii.
17. *Constituent Assembly Debates:* Official Report, Vol. VII, passim.
18. See my 'The Future of the Backward Classes' in A. Béteille, *Castes: Old and New*, Asia Publishing House, 1969.
19. P. V. Kane, *History of Dharamashastra*, Bhandarkar Oriental Research Institute (Poona), 1974, Vol. II, Part I, pp. 168–70.
20. Karnataka Backward Classes Commission, *Report*, p. 36; emphasis in original.
21. Fortes, *Kinship and the Social Order*, p. 147.
22. Durkheim's celebrated defence of individualism was made in the cause of Alfred Dreyfus. See É. Durkheim, 'Individualism and the Intellectuals' (trans. S. & J. Lukes) in *Political Studies*, Vol. XVII, 1969, pp. 114–30.
23. *Constituent Assembly Debates: Official Report,* Vol. VII, p. 39.
24. V. M. Dandekar and N. Rath, *Poverty in India*, Indian School of Political Economy (Bombay), 1971; T. N. Srinivasan and P. K. Bardhan (eds.), *Poverty and Income Distribution in India*, Statistical Publishing Society (Calcutta), 1974.
25. It is a noteworthy feature of sociological studies in India that they have paid hardly any attention to downward mobility as compared to the attention devoted to upward mobility. Again, this is partly because downward mobility is hardly visible any longer in the caste system, although it is very conspicuous in the class system.
26. N. K. Bose had pointed this out in studies he had made in the forties. See Bose, *The Structure of Hindu Society*, Chapter 11.
27. *Report of the Backward Classes Reservation Commission, Kerala, 1970*, Government of Kerala, 1971.
28. *A.I.R. 1963 S.C. 659*.
29. Reported in *The Statesman*, Delhi, 8 March 1980.
30. Bose, *The Structure of Hindu Society*, p. 132.
31. M. Weber, *The Religion of India*, The Free Press, 1958, p. 3.

Distributive Justice and Institutional Well-Being

The Problem

I would like to use this occasion to discuss the subject of reservation within a broad and general framework of enquiry. It has exercised the public mind for some time, and I am not unaware of the risks in attempting a scholarly treatment of problems to which so many events have brought such a pressing sense of urgency. But the subject is not only of immediate interest; it raises questions of great importance to social and legal theory that ought to be examined more fully and systematically than has been done so far. Moreover, I have worked on it for the last ten years, and I will try to develop here some of the arguments that were first presented in my Ambedkar lectures delivered in the University of Bombay in March 1980.[1]

Let me begin with a couple of observations on the public response to reservation in the last few months. It appears that those opposed to it were taken by surprise by the announcement in early August that the government would implement the recommendations of the Mandal Commission for reservation in employment. Yet those recommendations had been adopted by Parliament with unanimous acclaim eight years previously, on 11 August 1982. My own conclusion then was the somewhat fatalistic one that reservation as recommended by the Mandal Commission, or something similar, would be implemented sooner or later, since there seemed to be so little opposition to it from either politicians or intellectuals.[2] Perhaps those who were uneasy over the recommendations hoped that Parliament would, with the passage of time, forget its own unanimous approval and the whole issue would die down. I do not believe that the issue will die down even if the recent decision of the government is suspended or reversed for the time being.

A second striking feature of the public response is that moral passions have been raised to a very high pitch. Those opposed to reservation argue that injustice will be done by it to the 'forward' castes or that the rights of meritorious individuals will be sacrificed.

The V.T. Krishnamachari Memorial Lecture delivered at the Institute of Economic Growth on 12 November, 1990; also published in *Economic and Political Weekly*, Vol. XXVI, Nos. 11 and 12, March 1991.

On the other side, people argue that reservation will vindicate the rights of backward communities and remedy the injustices suffered by them for centuries if not millennia.

I would like to stress the extensive if not universal use of the language of justice and rights in the public debate on reservation. Those who are opposed to it tend to dwell more on the rights of individuals, whereas those in favour speak more of the rights of castes and communities. Affirmative action or reservation—including numerical quotas—has been a subject of public debate also in the United States. There the language of rights has figured far less prominently than here in arguments for and against reservation of the kind with which we are concerned. In America the proponents of even numerical quotas tend to make their case not in terms of justice but in terms of utility; not from arguments about rights but from arguments about policy.

When I compare the Indian case with the American, I sometimes wonder if we have a stronger sense of justice than the Americans; or if our sense of justice is a radically different one. That is not an easy question to answer, but I would like to keep it in mind as I go along. The *Dharmashastras*, by which Hindus were governed for many centuries, were animated by a very different sense of justice from the one on which the Indian Constitution is based. The old sense of justice corresponded to a distinctive culture and a distinctive social structure. It is far from clear to what extent it has been displaced by the changes taking place in Indian society and culture today.

The social and cultural changes of the last hundred years have been associated with the emergence and growth of administrative, educational, scientific, medical, financial and other institutions of a public or semi-public character that are either completely new or so different from their traditional counterparts as to count effectively as new. I would like to devote some attention to these institutions because many of our hopes and aspirations—whether we are for or against reservation—turn on their survival and success. They have a certain universal character in the sense that they are a part of the social landscape not only of modern India but of the whole modern world.

Universities, hospitals, laboratories, banks and bureaus have become an important part of the modernization of India. I introduce the term 'modernization' with some hesitation because it has acquired a somewhat bad odour through the persistent, though in my judgement misguided, exertions of social scientists with a radical vocabulary. But I need not discuss the concept further here, because there does not appear to be any systematic disagreement between supporters and opponents of reservation on the value to be assigned to a hospital, a science laboratory or a bank. Today the disagreement is not so much about the value of these modern institutions as about the criteria by which access to positions of respect and responsibility in them should be regulated.

Now, it is a sociological truism that all institutions do not work in the same way; each has its own rules and requirements, its own structure of rights and obligations, and its own internal culture. The structure of rights and obligations that govern a joint family would not be appropriate to a science laboratory; and the culture of a village *panchayat* would not be appropriate to a university.

I am not aware of any satisfactory definition of institutional well-being, and I do not attempt to provide one. It is easy to anticipate the difficulties that such an attempt must encounter. Yet, we have an intuitive sense of the well-being of individuals, and we can, at least for a start, use that sense to guide us when we speak of the well-being of institutions. It appears evident that institutions have conditions of well-being that differ from one type to another. These conditions can hardly be specified with the clarity and precision of a scientific formula. The best that we can do is to exercise our judgement in a sober and dispassionate manner on the basis of accumulated experience. At any rate, it is useful to remind ourselves that institutions cannot be squeezed and stretched at will without serious risk to their continued existence.

I have introduced the subject of justice, and I have made a few observations on institutions. I am now in a position to formulate my central problem in a provisional and tentative way: it is the problem of the compatibility between the requirements of institutional well-being and the claims made by or on behalf of disadvantaged groups in the name of distributive justice. How far can such claims be accommodated without damage to the interior lives of these institutions? I believe that it is disingenuous to pretend that every institution can accommodate without any strain all the claims that are made on it in the name of distributive justice.

Groups and Individuals

The fundamental issue in distributive justice is equality: a more equal or at least a less unequal distribution of 'the benefits and burdens of social co-operation'.[3] In that sense distributive justice seeks to go beyond equality in the purely formal sense: equality before the law, the equal protection of laws, or even formal equality of opportunity. Its central concern is, in the language of Rawls, 'to redress the bias of contingencies in the direction of equality'.[4] There can be no doubt that the concern for a more equal distribution of benefits and burdens figures in our Constitution, but I should point out that it figures most prominently in the part on Directive Principles of State Policy. I may also point out that some contemporary authors have questioned the very concept of 'distributive justice'[5], although I would like to leave open, at least for the time being, the question as to where any preconceived pattern of distribution is a matter of justice and where it is a matter of utility.

Any attempt to promote distributive justice must begin with a consideration of the existing inequalities in a society. The presence of large social inequalities is a striking feature of contemporary India. There are many different forms of inequality in Indian society, as in other contemporary societies. In our perspective, it is essential to keep in sight both inequalities between individuals and disparities between groups. Disparities between groups have been historically of great significance in Indian society, although they are not unique to it.

The distinction between individuals and groups is important in the present context since the claims made for a more just distribution of benefits and burdens on behalf of individuals are not identical with those made on behalf of groups. There are also broader historical and sociological reasons for paying attention to the distinction between the two. I would like to examine a little more closely the place assigned to individuals and to groups in different types of society and in different historical epochs.

Let us begin with the individual. While all societies are made up of individuals in the empirical sense, they do not all assign the same value to him or her as an autonomous moral agent.[6] Where a high value is placed on the individual, he is expected to make his own life for himself and to be judged on his own merit, irrespective of family or community. To be sure, this is only an ideal, but it is a social ideal that has acquired a peculiar force in the modern world. Alexis de Tocqueville had pointed out a hundred and fifty years ago that 'individualism' was a new expression to which a new idea—one may say a new ideal—had given birth.[7]

At least in the western historical experience, egalitarianism and individualism were closely linked in their origin.[8] Before the modern age, a different kind of social order prevailed in which the individual was subordinated to the group, and society was cast in a hierarchical mould. The eighteenth and nineteenth centuries witnessed not only large economic and political changes, but the displacement of the old social ideal of a hierarchy of estates by a new one of the equality of individuals as citizens. The new social ideal became anchored in new institutions, particularly in education and employment where equality of opportunity and 'careers open to talent' became the watchwords. Needless to say, these far-reaching changes in the institutional system did not lead at once to the elimination or even the reduction of inequalities in the distribution of income between individuals.

The subordination of the individual to the group is associated with the prevalence of birth over achievement: one's position in society is marked by the group into which one is born, not by what one achieves for oneself. Traditional India has been described as 'the land of the most inviolable organization by birth'.[9] Here the group prevailed over the individual more completely and more continuously than in any other society known to history. What counted socially was the village community, the joint family, and, above all, caste, but not the individual.[10] The individual had very little space for movement within society: he could of course fulfill himself outside society by adopting the way of the renouncer.

Whether in pre-revolutionary France or in traditional India—or for that matter in pre-modern China—the subordination of the individual to the group went hand in hand with the hierarchical arrangement of groups. The more rigid the social hierarchy, the more strict the subordination of the individual. If there be any sociological law, it is this: in all hierarchical societies—by which I mean societies that are hierarchical by design and not merely in fact—the individual counts for little and the group for a great deal; there is, so far as I know, no exception to this.

Whatever its other virtues—and it had positive functions in the traditional order—for a thousand years and more, caste has stood for the most rigid social hierarchy and at the same time for the most complete subordination of the individual to the group. However, things have not stood still in India for the last hundred and fifty years. The legitimacy of the traditional social hierarchy was shaken by the legal and economic changes introduced by the British in the nineteenth century. A first hesitant turn was taken in 1850 with the Removal of Caste Disabilities Act; a hundred years later, the Constitution of India not only questioned the legitimacy of caste, but repudiated it altogether.

The social reformers of the nineteenth century who sought to dismantle the edifice of caste did so from a mixture of motives and for a variety of reasons. They found the all-pervasive ranking of groups repugnant and even absurd in the light of the new social ideal of equality; but they also sought freedom for the individual from the restrictions imposed by the group, irrespective of its social rank. Today the situation has taken a different turn because opinion is sharply divided between those who believe that the cause of equality can be advanced under the banner of caste and those who believe that this is impossible.

In the traditional Indian village it was acknowledged that every caste had a just claim to a share—not an equal share, to be sure, but a certain share—of the social product. We may say that this reflected the traditional concept of distributive justice which attended directly to the claims of groups, and only indirectly and by implication to those of its individual members. The claim that the state should distribute the benefits of education and employment equitably between the different castes and communities is a strong one because it raises echoes of a social ideal that had prevailed in India for centuries.[11]

While it may be true that proposals for caste quotas in education and employment raise echoes of traditional ideas of distributive justice, this does not mean that nothing has changed between then and now. Although all castes could claim shares in the social product, it was clearly understood by all parties in the past that these shares must be unequal. Now the claim is for equal shares or shares proportionate to population. At a deeper level the caste system has changed fundamentally. The moral claims of castes over their individual members have weakened at all levels of society, and especially for the urban middle class where the battle over reservation is being fought. It will be safe to say that no caste today has the moral authority to enforce on its middle-class members any of its traditional sanctions. Having freed themselves from the moral authority of their caste, such individuals are now able to use it instrumentally for economic and political advantage.

In the traditional order, the village priest, or the village barber, or the village scavenger had a moral right to claim a share of the social product in the name of caste because each of them was bound by the moral authority of the caste of which he was a member. That moral authority has been, for good or evil, shattered for ever. On what ground can individuals now claim distributive shares for themselves in the name of their caste after having repudiated their moral obligations to it?

Rights and Policies

I would now like to turn to some of the specific arguments that have been made in support of reservations. I have to say, at the risk of appearing unpatriotic, that the best arguments of that kind have been made not in India but in the United States. It is true that the American situation is in many ways different from the Indian, but there too preferential policies in favour of disadvantaged groups defined by race and ethnicity—rather than caste and tribe as in our case—have become a subject of public debate.

To my mind the most forceful and at the same time the most acute argument in support of reverse discrimination—including numerical quotas—has been made by the American lawyer Ronald Dworkin who is Professor of Jurisprudence in Oxford. The current bias of intellectual opinion in the United States is much more in favour of equality of opportunity and much more against discrimination in any form than in India, even though it is the Indian Constitution, and not the American, that has an equal opportunities clause and an anti-discrimination clause in its part on Fundamental Rights. Given that bias of opinion, those who argue for reverse discrimination in the United States have to present strong arguments.

Dworkin first presented his argument for reverse discrimination in connection with the DeFunis case in a paper that received wide attention when it appeared in 1977 in *The New York Review of Books*.[12] A white man named Marco DeFunis had applied for admission to the Law School in the University of Washington, but was rejected even though he had done better on college grades and test scores than others belonging to disadvantaged races who were admitted. DeFunis was by conventional academic criteria the better candidate, but he was passed over in order to make room for others, ostensibly on grounds of race. The case led to a sharp division of liberal opinion in the United States. Dworkin supported the decision of the Washington University Law School to reject DeFunis in the interest of a more racially-mixed student body.

Those who are opposed to reservation on the basis of race, caste or ethnicity come forward, sooner or later, with the argument that it entails a sacrifice of individual merit. Dworkin faced that argument directly by saying that merit did not by itself create the kind of right that the Law School was accused of violating by denying admission to DeFunis. He did not say that merit was irrelevant or unimportant, only that it did not create a right of admission in DeFunis or anyone.

Dworkin maintained that the argument that the Washington University Law School had violated an individual right by denying admission to DeFunis was misconceived, because he did not have such a right in the first place. 'DeFunis plainly has no Constitutional right that the state provide him a legal education of a certain quality'.[13] Dworkin was not saying that the individual had no rights; he was saying that he did not have the kind of right that was being claimed by DeFunis or on his behalf. 'DeFunis does not have a right to equal treatment in the assignment of law school places; he does not have a right to a place just because others are given places. Individuals may have a right to equal treatment in elementary education, because someone who is denied elementary education is unlikely to lead a useful life. But legal education is not so vital that everyone has an equal right to it'.[14]

Dworkin returned to the same argument a few years later in an eassy on the Bakke case where a white applicant had been denied admission to a medical school that had set aside a number of places for members of 'educationally and economically disadvantaged minorities'. He repeated the argument that Bakke had no constitutional right that had been violated by the medical school when it denied him a place in the interest of its affirmative action programme. That programme was a good one because it served a useful policy, and, although it might cause disappointment or even hardship to the individual, it did not violate his constitutional rights.[15]

I am a little uncertain as to how fully these arguments apply in the Indian case. The Indian Constitution, unlike the American, has clear provisions proscribing discrimination and prescribing equality of opportunity in the part on Fundamental Rights. Further, it is the citizen as an individual, rather than any caste or community, who has the right to equal opportunity. It is true that the right is not absolute or unqualified since it has to accommodate special provisions; but that accommodation cannot be so extensive as to render the right fictitious.

Dr Ambedkar had, in the Consituent Assembly, drawn attention to this kind of threat to equality of opportunity. He had said,

> Supposing, for instance, reservations were made for a community or a collection of communities, the total of which came to something like 70 per cent of the total posts under the state and only 30 per cent are retained as the unreserved. Could anybody say that the reservation of 30 per cent as open to general competition would be satisfactory from the point of view of giving effect to the first principle, that there should be equality of opportunity?[16]

Whether or not the individual citizen has, on the ground of 'merit', an unqualified right of admission to a medical college or the civil service, the principle of equality of opportunity is a first principle that we cannot afford to devalue. It cannot but be devalued by the extension of massive caste quotas into every area of public life.

My purpose in drawing attention to the American debate was to make the point that quotas for disadvantaged groups are best viewed as matters not of right but of policy. In the United States, the strongest arguments in support of reverse discrimination are made not on grounds of rights and justice but on those of policy and utility. Dworkin rejects categorically the assumption that 'racial and ethnic groups are entitled to proportionate shares of opportunities', and adds, 'That is a plain mistake; the programmes are not based on the idea that those who are aided are entitled to aid, but only on the strategic hypothesis that helping them is now an effective way of attacking a national problem'.[17] Among other things, this allows a degree of freedom and flexibility in the formulation and administration of such programmes.

In India we face a somewhat different situation. The case for reverse discrimination is made persistently, and with increasing intensity, in the language of rights. This at once raises the temperature of the debate and forces people to adopt intransigent positions. Understandably, they find it far more difficult to yield on what they believe, or are led to believe, to be matters of right and justice than they would on matters of utility or policy.

It is difficult to see how the idea that castes and communities have *rights* to proportionate shares in public employment can be made compatible with the working of a modern society committed to economic development and liberal democracy. It is true that caste continues to operate in many spheres of social life; but it does not do so any longer as a matter of right. The continued existence of caste is one thing; its legitimacy is a different thing altogether. The attempt to invest the caste system with legitimacy by claiming that its constituent units have rights and entitlements is bound to be defeated in the end; but in the mean time it can cause enormous harm to society and its institutions.

The persistent use of the language of rights in the public debate for and against reservations is bound to lead to an increase in the consciousness of caste, and in that way to defeat the basic objective of affirmative action which is to reduce and not increase caste consciousness. All parties to the debate say that they wish to dismantle the structure of caste. But caste is not a material edifice that can be physically dismantled and destroyed. It exists above all in the consciousness of people—in their deep sense of division and separation on the one hand and of rank and inequality on the other. How can we exorcise caste from the public mind by deepening the sense in society that castes are entitled to their separate shares as a matter of right?

Undoubtedly, there are vast disparities in Indian society that need to be redressed. For that we require policies that will be useful and effective. Flexibility is of the essence in the design and application of policies to redress disparities that have arisen from many causes and not just caste. The insistence on rights where none exist in either law or morality destroys the very flexibility without which no policy of affirmative action can be useful or effective.

Individuals and Institutions

I take it for granted that policies for the redress of severe social and economic disadvantages are in themselves desirable. Such policies have to aim at different sectors of society and at the widest possible base. An obvious field for the application of preferential policies is that of education where the maximum attention should be devoted to primary and secondary education which develop the base on which the success of higher education depends. Other fields for which preferential policies may be designed include those of child-care, health and housing.

Policies, unlike rights, are not absolutes; they have to be examined in terms of costs and benefits. We may not always be able to measure these, but that should not prevent us from trying to form clear judgements about them. Both costs and benefits must be taken into account in assessing any policy of affirmative action. Here I shall confine myself to job reservation in public institutions. This is not because it constitutes the most useful or effective application of the principle of redress, but because it helps to bring into focus issues that I consider important and because it has received such wide public attention.

When people think of reservations today, they think first and foremost of employment, and, above all, of salaried employment in the services of the union and state governments and in other public institutions. The conditions of success and failure in securing such employment have become a kind of acid test of the fairness of the system among the supporters as well as the opponents of reservation. This is inevitable in an economy where there is so much unemployment all around and where for every vacancy there are numerous applicants. Salaried employment is a source of security and status to the individual and his family; but it is difficult to judge how far the benefits of public employment can or should spread from the individual to his kith and kin, and to his community.

The middle class, or what some now describe as the service class, looks after its own, and tends to reproduce itself from one generation to the next.[18] In every society, a crucial part is played in this process by the family. It is true that the caste system is a serious obstacle to equality of opportunity; but at a deeper level the family system is a far more persistent obstacle to it. I make this point because, although the family system is universally present, it is stronger in some societies than in others. Given the exceptional strength of the family among all social classes in India, all talk about equality of result as an attainable objective has to be taken with a very large pinch of salt. Thus, the first institutional obstacle that the removal of inequality encounters is the family: it is not an easily removable obstacle.[19]

There is of course a certain amount of mobility or circulation of individuals between different social strata, despite the persistence and strength of the family. The extent of such mobility varies from one society to another, and social policy has undoubtedly some part to play in increasing mobility by removing obsolete or artificial obstacles to it. In India the tendency of the service class to reproduce itself is in both appearance and reality particularly marked. Firstly, because this class is so small relatively to the population as a whole, its unity and continuity appear stronger than in societies where it is relatively large.[20] Second, the dramatic changes in the occupational structure of the advanced industrial societies have led to a net increase of upward over downward mobility;[21] in the absence of such changes in India, there has been little, if any, aggregate increase in upward mobility. Thirdly, the residues of a rigidly hierarchical system have acted as obstacles to the free flow of lower-caste individuals into the service class.

There is no denying the fact that individuals from different castes are found in very different proportions in the service class. There are good grounds for feeling that such wide disparities are undesirable, and reasons for believing that they can be corrected, at least to some extent, by useful and effective policy. But we must first understand the reasons why the lower castes are so thinly represented in the higher occupations. There is extensive prejudice against their members, but this applies particularly to the Harijans and Adivasis. There is also a marked decline in the number of qualified candidates as we move down from the higher to the lower levels of the caste hierarchy; here the family system plays a crucial role in socializing children differently and transmitting cultural capital to them unequally. The best course would be to make all-out efforts to expand the pool of qualified candidates at the lowest level; but that would take the kind of patience and care that no government in India has so far shown. A quicker course, whose effects would show immediately in official statistics, would be to alter the proportions directly through the reservation of jobs.

Changing the pattern of employment through extensive reservations in public institutions will affect some individuals favourably, but others adversely. It may be argued, quite plausibly, that even if the gains to some individuals are balanced by the losses to others, there may be a net benefit to society, certain conditions being met, by having a better mix of public servants. Upper caste candidates will face some reduction in their opportunities for employment, but it does not necessarily follow that they can claim that their rights have been violated. There will no doubt be disappointment and discontent, and the state will have to take account of them, but as matters of policy rather than right.

The government has in fact responded with some sympathy and concern for those whose employment opportunities are likely to be reduced on account of reservations. It has offered to create more employment so that reservations do not dramatically reduce the number of jobs available through open competition. I leave it to the experts to determine the capacity of the economy to absorb the additional employment productively. I presume that much will depend on the scale on which these shifts are made and the time span over which they are spread, but that from the viewpoint of employment there is nothing in principle against creating some new jobs for open competition while increasing the number of those in the reserved category.

Any discussion of changes in patterns of employment must give due importance to magnitudes, and it has to be admitted that the magnitudes involved here are relatively small. We are dealing in fact with jobs that have to be counted in tens of thousands or at best in lakhs in a population of 850 million persons or around 170 million households. In these circumstances, job reservation can hardly be expected to bring about a significant reduction in social and economic disparities by altering the balance of employment. It is in this sense that Mr Mandal's Commission has characterized them as 'palliatives'.

Let me repeat that I do not wish to make light of the problem of unemployment. That is a problem of the first magnitude in both rural and urban India which calls for the most serious attention in its own right. Here I wish to make the simple point that altering the caste composition of the service class, with or without some addition of posts in the services of the union and state governments, cannot possibly be regarded as a viable policy for solving the problem of unemployment in an economy that is predominantly rural, predominantly agricultural, and predominantly outside the organized sector.

A consideration of employment and the opportunities gained and lost by individuals of different castes gives us one view of reservation. We get a different, and in my opinion a more significant, view of it by a consideration of the nature and functioning of institutions. In the first view, we deal with quantities, and the quantities, as we have seen, are not very large in relation to the labour force as a whole. In the second view, the number of individuals who gain or lose is less important than the quality of each of the institutions concerned.

In considering institutions, we have first of all to keep in mind the differences in their nature and functioning. Their size, or the number of persons employed in them, is of less significance. The Supreme Court is an institution of first importance; we do not take the number of jobs available in it as the first consideration in judging its quality and efficacy.

I would like to stress the very great need to maintain a differentiated approach to institutions in the matter of reservations. That approach is clear in the Constitution which has mandatory provisions for reservations in the Lok Sabha but not in the Supreme Court. I have sometimes heard it said that this is because the Supreme Court has to maintain a level of excellence that we do not ordinarily expect from the Lok Sabha. That is a mistaken idea. The two institutions differ not in the levels of excellence expected of them but in their functions. The Lok Sabha performs representational functions of a kind that the Supreme Court does not.

A certain conception of democracy has come to prevail among large sections of our society, including the intelligentsia, that can only be described as the populist conception. That conception requires that all public institutions, irrespective of nature and function, should involve the representation and participation of the people; it is as if all public institutions should be and act like political councils and committees. What is surprising is not so much the eagerness of the politicians to impose the political model as the general model for public life everywhere, as the readiness of the intellectuals to accept it in the name of democracy. I hardly need to say that that is the opposite of my conception of democracy which assigns a central place to institutional autonomy based on an understanding of and a respect for the differentiated character of institutions.

I take it for granted that courts, hospitals, universities, laboratories and banks are useful not just to the people to whom they provide employment but for the public at large and for society as a whole. The social utility of a public institution has to be judged not just by the criterion of employment but by a whole range of criteria among which employment need not be the most important. Further, it is neither the purpose nor the function of these institutions to provide representation to the different sections of Indian society but to meet other requirements that differ widely from one type of institution to another. It would be unreasonable to expect the High Court of Karnataka to function in the same way as the Indian Institute of Science, or the recruitment of scientists to be made by the same criteria that are used in the recruitment of judges.

It is good not only that institutions should be differentiated from each other, but also, though for a different reason, that each should be as varied in composition as possible. An institution that is all-male or all-Bengali, or has only Hindus or only Brahmins is likely to be less resilient, less sensitive and less rich in the quality of its life than one with a more mixed composition. It is a fact that, as a result of complex social and historical processes, the Supreme Court did not have until the other day any woman member on the bench. It might have been better if we had had from the start more women judges, not because women have a right to their proportionate share on the bench but because their presence might have widened the range of experience of its members. However, that cannot become even now an overriding consideration in constituting the Supreme Court bench or any other bench.

What is true of courts of justice is true to a greater or lesser extent of other institutions as well. It is certainly true to a large extent for the university though perhaps not to the same extent for a specialized research institution. A university stands for the meeting of minds, and variety in the social composition of its faculty and student body is definitely an asset in that regard. Where a university has among its members individuals from severely disadvantaged groups, their experiences and perspectives add to the variety and richness of its life. I view this not as a matter of justice or rights, but as one of institutional well-being.

But an institution cannot enhance its well-being by compromising the ends and means specific to it merely for the sake of greater variety. Nor is it true that mere variety of social background is a guarantee of equal variety in the thoughts and feelings of people. There is little reason to believe that those who move into privileged positions from severely disadvantaged backgrounds always remain faithful to the thoughts and feelings of their early environment; in many cases, if not most, they try instead to suppress those thoughts and feelings, to escape from their past in order to adjust more effectively to the demands of their new ambitions. Rapid upward mobility affects peoples' perceptions and orientations in ways that still remain largely obscure.

It is said that a teacher or a doctor or a lawyer has to deal with problems not only on a technical plane but also on the human plane. But what does it mean for a teacher to deal with his students—or a doctor with his patients—on the human plane? It means that the teacher should be able to put himself in the place of his student—or the doctor in the place of his patient—and view his or her problems with concern and sympathy, and not just technical ability. No university or hospital or court of justice could function as an institution unless its responsible members had some capacity to put themselves in the places of others, without regard for caste, creed and provenance.

Every college or university department has to deal with students who have special problems that arise from social disadvantages or personal misfortunes of a hundred different kinds. Some teachers deal with such problems more successfully than others. It cannot be said that problems that arise among students from personal misfortune or social disadvantage can be dealt with only by teachers who have themselves experienced identical problems. Nor can it be said that the only problems among students—or patients, or some other category of citizens—that call for treatment with special concern and sympathy are those that arise from the disadvantages of caste. Nothing can be more misleading than to argue that all the problems that we, as members of institutions, have to deal with on the human plane arise from the past excesses of caste, and from those alone.

Institutions of the kind I have in mind have not only to deal with problems on the human plane, they have also to deal with them impersonally. There has to be a balance between the requirements of treating individuals with concern and sympathy, and treating them without fear or favour; but the right balance will be different for different institutions. Institutions devoted to science and scholarship have greater flexibility than purely administrative institutions where conduct is, or ought to be, bound more strictly by impersonal rules. But no modern institution can free itself fully from the demands of the latter. However acute the problems faced by a particular student, the teacher has to apply the same standards in checking his experiments or marking his examination papers that he applies to others. Nothing can be more misconceived than to condone faulty experiments or wrong calculations on 'humanitarian' grounds.

I believe that it is here, in the domain that is, or ought to be, governed by impersonal rules that our modern institutions will face their most severe test. Some years ago I had made a distinction between societies that are governed by rules and those that are governed by persons.[22] That was a crude distinction and defective on a number of points. It is nevertheless the case that in our traditional institutions—village, caste and joint family—personal considerations prevailed to a large extent, although this does not mean that all persons received equal treatment or even equal consideration. Modern instiutions are organized differently, reflecting partly a change in scale and partly a shift in normative orientation. Their organization requires the prevalence of impersonal rules over claims based on ascribed social positions such as those of kinship, caste and community.

The intrusion of ascriptive criteria, or considerations of community, caste and kinship into institutions that value performance and achievement vitiates not only their composition but also their functioning. Two factors are very important here: (i) the scale on which these ascriptive criteria are introduced, and (ii) the legitimacy accorded to them.

So long as only a few places are kept aside in order to create special opportunities for members of severely disadvantaged groups such as the Harijans and the Adivasis, considerations of caste and community can be kept under control and not allowed to vitiate the functioning of institutions. But those very considerations will be bound to loom large where half the places in an institution are set apart for specified castes and communities and the other half filled by open competition.

The question of legitimacy is also important. It is said that in India, institutions such as universities, hospitals and even research laboratories are already riddled with caste. So far at least, people say this not to praise these institutions but to blame them. But how can we blame public institutions for being caste-ridden if we declare that castes and communities are entitled to their 'due share' of positions in them as a matter of right? I go back to the crucial importance of the distinction I made earlier between matters of right and matters of policy.

To me the most important argument in support of caste quotas is not the one about employment but the one that it is only through them that the interests of backward communities can be protected in public institutions. This is not a new argument, but one that was first formulated by the British on the basis of their perception of the Indian national character. British civil servants widely believed—and sometimes said—that Indians could not be trusted to deal justly, or fairly, or even-handedly with other Indians if they were of a different caste or community. They believed, perhaps sincerely, that Indians lacked the character to act without fear or favour, that they had a strong conception of the interests of their caste or community, but none, or only a weak one, of the public interest. That is a severe judgement, but we must not flinch from facing it.

Three generations of Indian nationalists sought to disprove the British view of the Indian character, and their spirit animates the eloquent letter to the President with which Kaka Kalelkar forwarded the Report of the first Backward Classes Commission.[23] What could be more ironical than the determination of our present political leaders to prove, after four decades of independence, that the British perception of the Indian character was right after all?

The argument that job reservation is essential because we need watchdogs in every public institution to look after the interests of Harijans, Adivasis, backward castes and minorities needs to be brought out more fully into the open. If the argument is right, or, despite being wrong, is widely believed to be right, our institutions cannot function as they ought to: their well-being will be irreparably damaged. We cannot say what kinds of universities, hospitals and scientific laboratories we will then have in the future, but they will be very different from the ones that at the time of independence people had hoped for.

It is said that those institutions from which we expect the highest levels of excellence in terms of international standards should be exempted from the rule of reservations so that they are able to attract the very best talent. While that may be so, I do not believe that excellence in that sense is the first consideration in institutional well-being. There are other, more important considerations, such as those of probity, integrity and trust. They are first considerations because, without them, institutional, as against purely individual, excellence will have very little meaning or content.

Every modern institution has a framework of more or less formal rules that define the rights and obligations of its individual members and specify sanctions to uphold them. But its well-being depends on much more than a framework of formal rules. No hospital or university or bank could operate successfully if its members sought to assert their formal rights or to have their obligations formally specified at every turn. These rights and obligations have to be so internalized as to enable most persons to take them for granted most of the time. There has, thus, to be a fiduciary component, or a component of trust, at the very core of such an institution as of every social institution. It is this fiduciary component that is put seriously in question when people claim that caste biases cannot be corrected without the representation of all major castes or groups of castes in hospitals, laboratories, universities, banks and bureaus.

Apart from its structure of rights and obligations, an institution has also its own subculture, consisting of a distinctive set of ideas, beliefs and values. This subculture varies from one type of institution to another, but there are also certain commonalities among modern institutions in general. Whatever the difference from one subtype to another, the subculture of a modern institution—its spirit or ethos—is at the opposite pole from the culture of caste. If it is to function properly, and not necessarily at a high level of excellence, a hospital, a laboratory or a secretariat must be to some extent insulated from that culture. Here again, it is no argument to say that these institutions are already to some extent infected by caste; it cannot be sound policy to make that infection as deep and extensive as possible.

It is narrow and short-sighted to regard the well-being of institutions as of concern to only those to whom they provide a livelihood; to judge them by their capacity to meet the demands of employment is to do precisely that. An institution such as a university or a hospital or a bank has responsibilities not only to its own members but also to a much wider public. When a public institution suffers decline due to faults in the system of recruitment and rewards, the resulting harm affects not only its internal order but also, and necessarily, its capacity to fulfill its obligations to society as a whole.

India is a large, complex and changing society that provides ample scope for combining the rhetoric of social justice with the pursuit of private interest. In several crucial spheres, including health and education, both public and private facilities are available. Those who have the resources can and do make use of both, and the members of the service class have shown themselves to be increasingly ambidextrous. Since access to private facilities requires both money and influence, the poor and disadvantaged have to depend on public institutions irrespective of their quality. When the influential members of a society cease to make use of a public institution because it is under decline, its further decline becomes inevitable; it then survives only as the refuge of those who cannot turn elsewhere. The neglect of public institutions hits harder the poor and the disadvantaged because for them there is no alternative; therefore, it is not at all obvious that the extension of caste quotas will reduce and not increase inequality overall.

It is not that the intelligentsia are unaware of the corrosive influence of caste on the kinds of institutions I have spoken about. If some of them are nevertheless prepared to see the corrosion spread a little further, they have a number of arguments on their side. An extreme position is that these are basically bourgeois institutions of little intrinsic value, and have to be overhauled to make room for alternative institutions to be constituted on Marxian, or Gandhian, or some other principles. This is too large a subject to enter at this point; and it is difficult to discuss it in the absence of any clear account of what such alternative institutions would look like or how they would function.

Others feel that these institutions are of intrinsic value, and that they should be preserved and improved, but also that they can be made to bear an extra burden, at least for some time, in the interest of greater equality overall. That is not an unreasonable view, and it deserves serious consideration. Much depends on how large the extra burden is likely to be, and how we weigh it against the presumed benefits. Even if we take an optimistic view of those benefits, which I do not, we have to consider the costs at two levels: firstly, the costs to the institutions severally; and, secondly, the cost to society as a whole of concentrating the extra burden so heavily on a few strategic institutions in the belief that the citadels of elitism must yield before the demands of distributive justice.

State and Civil Society

Once the uneven distribution of castes in public institutions comes to be perceived and represented as a problem in distributive justice, institutional well-being takes a back seat. This does not mean that institutions will be consciously or willfully harmed. But their requirements will be ignored, and the costs to them from ambitious but ill-conceived policies to attain equality and justice will receive little or no attention.

It is in a way natural that after having lived in a hierarchical society for centuries, Indians should now be eager to establish a new pattern of distributive justice here and now. The old ideal has lost its appeal and is being replaced by a new one. But everything does not change when people decide to change their ideal from hierarchy to equality. The passage from a hierarchical society to one based on equality of status and opportunity has been a slow and painful one in the western world, and nowhere has it led—or can it lead—to the elimination of inequality in every form.

We in India have barely begun the passage from a hierarchical to an egalitarian society. It is true that we have changed our laws and our Constitution, but many other things remain unchanged. Many of the old material conditions of life remain, and many ideas are still cast in the same hierarchical mould as before. There is widespread poverty, illiteracy, ignorance, superstititon and prejudice. The economic forces that loosened the hierarchical structures of western society in the eighteenth and nineteenth centuries have had little scope for expansion in India, and the impact of education has been limited. Yet we feel that what has failed on so many fronts can somehow be made good by powerful and effective governmental action.

It is a mistake to believe, as many tend to do under frustration or despair, that every desirable state can be brought into existence by the government. In India, paradoxically, the belief in the power and efficacy of government as such has increased with the experience of corruption and inefficiency of every successive government. It is one thing to have a sound policy for education or employment, but quite another to design one that can abolish hierarchy and inequality, or establish equality of result.

It is not true that the natural advance of equality is always helped and never harmed by the increasing penetration of public life by the government. That the government is always for the people because it has declared itself against capitalism, elitism and patriarchy is a myth that can no longer be used to give any direction to policy. The machinery of government is an independent source of inequality in all societies, and can be particularly oppressive in those that have a predominantly rural population with very low levels of income and education. Undoubtedly, government can do something to remedy these very conditions, but it cannot do everything; where it tries to do too many things, it expands its own apparatus to the detriment of the public interest.

The fact that some inequalities can be removed or reduced by direct governmental action does not mean that they can all be removed or reduced by it. Here I would like to draw attention to the distinction between (i) the removal of disabilities, and (ii) the equalization of life chances. The two are by no means unrelated; but they are not the same in terms of either priority or feasibility, and they call for different types of strategic action.

Any programme for the advance of equality must give first priority to the removal of the disabilities imposed by law and custom.

Landmarks in this direction were Article 17 of the Constitution abolishing untouchability, and the adoption in 1955 of the Untouchability (Offences) Act, amended and renamed in 1976 as the Protection of Civil Rights Act. In the traditional order, disabilities were imposed in their most severe form on the untouchables, but the order as a whole was based on privileges and disabilities that were upheld by both *shastric* and customary law. Hence, the removal of disabilities has to be viewed as the first condition of the change from hierarchy to equality.

Experience has shown that disabilities that have been maintained by law and custom for centuries cannot be effectively removed by a single act of legislation. Bad laws may be annulled by the state, it is far more difficult to legislate effectively against bad customs. It is one thing to legislate equal rights for all citizens but quite another to make those rights secure for those who have been excluded from civil life for generations. There is much scope for affirmative action by the government to give security to the rights made available in principle to all. Although the problem takes a particularly acute form for the Harijans and requires special measures in their case, it is a general one for the vast masses of poor and illiterate persons, to some extent irrespective of caste. It may be useful to have some Harijan officers to keep an eye against violations of the Protection of Civil Rights Act. But reservation of posts beyond a point becomes counter-productive when it creates or reinforces the feeling that the rights of the weak can be protected only by those of their own caste.

The removal of disabilities is not only an urgent task, but, at least in principle, also a feasible one. The question of the equalization of life chances is altogether different. The former may be said to pertain to the legal and the latter to the economic domains, broadly conceived. We know from the experiences of other countries that started before us that legal equality and economic equality do not advance in the same rhythm. The experience of the western countries was that inequality of income was increasing during roughly the same period which experienced a steady advance of legal equality.[24] It is difficult to guarantee that, even with the best economic management, things can be made to turn out differently in India.

Therefore, we must not allow ourselves to be diverted by the declaration that what we should have in India is not just equality of treatment or even equality of opportunity, but equality of result.[25] It can hardly be said that by imposing caste quotas in government employment we are taking a significant step towards equality of result—or any kind of equality—in a country where most people remain ill-fed and unlettered. The rhetoric about equality of result does very little good; if taken seriously, it can cause much harm to the operation of economic forces and social institutions.

The record of economic management in India since independence has been uneven; it has been better than many would make it out to be, but it has not been one of spectacular or unqualified success. Economic management has to be based on a sober assessment of feasibilities. Where it has failed in India, an important cause of failure has been the pursuit of unattainable objectives. Providing basic facilities for health, childcare and education are slow and difficult, but feasible ways of bringing about some equalization of life chances. Providing 'good jobs' for all and establishing equality of result are not attainable objectives.

Faith in the capacity of the state to abolish the class system, or the disparities between groups, or to otherwise transform the basic structure of society by direct intervention in its processes has not been equally marked everywhere or in all historical periods. It has been particularly marked in some countries, including India, in the period through which we have lived. There are reasons to believe that that historical period is in some significant sense drawing to a close. We live in an age in which our consciousness is formed not only by our own particular experience but by the historical experience of the world as a whole. In reviewing our prospects for the future, it is important to take some account of that historical experience.

The Bolshevik revolution not only created a new type of society—a new civilization, as some would call it—but extended the horizon of possibilities in the minds of people in many countries. The Chinese revolution of 1949 was another major step in a new direction. Our own concepts of 'planned democracy' and 'participatory democracy' have been indelibly marked by these experiences.

Indian ideas about planning and a centrally-regulated economy and society have been shaped significantly by the perception of their success in the Soviet Union and other East European countries. I am talking now not of planning as a merely technical exercise in economic management, but of a whole outlook and philosophy for social and economic change. Many things went wrong with our planning; but these were viewed as technical failures that could be corrected by better techniques and bigger plans. Our perception of the overall success of planned social change in the U.S.S.R. remained as a perenniel source of assurance that ultimately our plans too would meet with success. The state grew socially more ineffective as it tried to increase its power; but the intelligentsia continued in the belief that all this would be corrected and that India too would have, as some other countries seemed to have, a socially effective state.

A massive current of change is now sweeping through the whole of eastern and central Europe, across those countries that have stood for socialism and the planned society. Some good things are bound to be swept away by it along with many that are obsolete, ineffective and oppressive. It is too early now to attempt a balance sheet for the new arrangements that are replacing the old ones. But one thing is clear beyond any doubt: in Poland, in Czechoslovakia, in Hungary, in East Germany, and, above all, in the Soviet Union, a new consciousness has come into being that makes it impossible to view the relationship between the state and civil society in the old way. We will only jeopardize our own future by insulating ourselves from this new consciousness with the protective armour of old shibboleths and outmoded slogans.

NOTES AND REFERENCES

1. A. Béteille, *The Backward Classes and the New Social Order*, Delhi: Oxford University Press, 1981; republished, along with other essays on related topics, in A. Béteille, *The Idea of Natural Inequality and Other Essays*, Delhi: Oxford University Press, 2nd edn., 1987. See, pp. 1–44.
2. A. Béteille, 'The Indian Road to Equality', *The Times of India*, 28 August 1982. See, pp. 80–4.
3. J. Rawls, *A Theory of Justice*, London: Oxford University Press, 1927, p. 4.
4. Ibid., pp. 100–1.
5. F. A. Hayek, *Law, Legislation and Liberty*, vol.2, *The Mirage of Social Justice*, Chicago: University of Chicago Press, 1976.
6. L. Dumont, *From Mandeville to Marx*, Chicago: University of Chicago Press, 1977; A. Béteille, *Individualism and the Persistence of Collective Identities*, Colchester: University of Essex, 1985.
7. A. de Tocqueville, *Democracy in America*, New York: Knopf, 1956, vol. 2, p. 98.
8. A. Béteille, 'Individualism and Equality', *Current Anthropology*, vol. 27, no. 2, April 1986, pp. 121–34.
9. M. Weber, *The Religion of India*, Glencoe: The Free Press, 1958, p. 3.

10. Jawaharlal Nehru wrote about the structure of traditional Indian society; 'This structure was based on three concepts: the autonomous village community, caste, and the joint family system. In all these three it is the group that counts; the individual has a secondary place' (*The Discovery of India*, Bombay: Asia Publishing House, 1961, pp. 247–8).
11. N. K. Bose, *The Structure of Hindu Society*, Delhi: Orient Longman, 1975.
12. R. Dworkin, *Taking Rights Seriously*, London: Duckworth, 1984, pp. 223–39.
13. Ibid., p. 225.
14. Ibid., p. 227.
15. R. Dworkin, *A Matter of Principle*, Cambridge, Mass.: Harvard University Press, 1985, pp. 293–303; for a slightly different position, see A. Béteille, 'Equality as a Right and as a Policy', *LSE Quarterly*, vol. 1, no. 1, 1987, pp. 75–98.
16. *Constituent Assembly Debates: Official Report*, vol. vii, p. 701.
17. Dworkin, *A Matter of Principle*, n.15, p. 297; see also T. Nagel, 'Equal Treatment and Compensatory Discrimination' in M. Cohen, T. Nagel and T. Scanlon (eds), *Equality and Preferential Treatment*, Princeton: Princeton University Press, 1977, pp. 3–18.
18. P. Bourdieu and J.-C. Passeron, *Reproduction in Education, Society and Culture*, Beverley Hills: Sage, 1977; also P. Bourdieu, *Distinction*, Cambridge, Mass.: Harvard University Press, 1984.
19. A Beteille, 'The Reproduction of Inequality: Occupation, Caste and Family', *Contributions to Indian Sociology* (n.s.), vol. 25, no. 1, 1991.
20. V. Subramaniam, 'Representative Bureaucracy: A Reassessment', *American Political Science Review*, vol. 61., no. 4, 1967.
21. J. H. Goldthorpe, *Social Mobility and Class Structure in Modern Britain*, Oxford: Clarendon Press, 1980.
22. A. Béteille, 'Rules and Persons', *The Times of India*, 4 November 1968, elaborated in A. Béteille, 'The Social Framework of Agriculture' in L. Lefeber and M. Datta-Chaudhuri (eds.), *Regional Development, Experiences and Prospects*, The Hague: Mouton, 1971.
23. Government of India, *Report of the Backward Classes Commission* (Kalelkar Commission), Delhi: Manager of Publications, 1956, vol. 1, pp. i–xxxiii.
24. S. Kuznets, 'Economic Growth and Income Inequality', *American Economic Review*, vol. 45, no. 1, 1955.
25. 'Equality of results', according to Mr Mandal, is 'the real acid test of effective equality'. Government of India, *Report of the Backward Classes Commission* (Mandal Commission), New Delhi: Controller of Publications, 1980, 1st part, p. 22.

Appendices

I

Reservations: The Problem

Any discussion of reservation for the backward classes must keep in view two important distinctions. The first is the distinction among the various spheres of public life: the case for reserved seats in bodies such as Parliament and the Assemblies to which members are elected is different from the case for reserved posts in bodies such as the civil service to which persons are appointed. The second is the distinction between the Scheduled Castes and the Scheduled Tribes on the one hand, and the Other Backward Classes on the other; the moral basis of the claims for special treatment of the Harijans and the Adivasis is quite different from the moral basis of the claims made by the various castes and communities which seek inclusion among the Other Backward Classes.

Seminar, No. 268 (Reservations), December 1981.

Any policy of reservation actually adopted may fall between two extremes. At one extreme will be a policy of reservation for only the Scheduled Castes and the Scheduled Tribes, and only in bodies constituted by election such as Parliament, the Assemblies, etc.. At the other extreme will be a policy of quotas not only for the Scheduled Castes and the Scheduled Tribes but for a whole range of other communities as well; and quotas not only in the central and state legislatures, but also in the civil and military services, in universities and colleges, in hospitals and laboratories, and in other public institutions. Some states in South India, such as Karnataka and Tamilnadu, seem to be moving towards a comprehensive system of quotas in virtually every sphere of public life.

There are within the very objectives of reservation contradictions that can no longer be swept under the carpet. The first Commissioner for the Scheduled Castes and the Scheduled Tribes, L.M. Shrikant, had pointed out that, instead of bringing the backward classes closer to the rest of society, reservation was serving to harden and perpetuate the identities of the Harijans and the Adivasis. The beneficiaries of reservation might move into and strengthen a new area of Indian society in which the individual is known by his occupation and no longer by his caste; or, they might seek to articulate the interests of their community, in which case the identities of caste and tribe will be strengthened and not weakened. Policy makers must make up their minds as to what they want out of reservation: they might want it either for the promotion of individual mobility or for the protection of collective interests, but the two objectives cannot be combined beyond a point.

Any attempt to assess the benefits of reservation must consider its benefits to individual members of the backward classes separately from its benefits to the backward classes considered as a whole. We have no reason to believe that whenever an individual Okkaliga or an individual Padayachi benefits from job reservation, the benefits are in any significant sense shared by all Okkaligas or all Padayachis. In point of fact, the jobs that can be reserved are so very few in comparison to the population of the backward classes (whether in the broad or the narrow sense) that it would be absurd to expect any significant redistribution of income among castes to follow directly from job reservation.

If the benefits of reservation are difficult to measure, its costs are no less difficult to determine. In any case, we need a far more differentiated awareness of the costs of reservation than we have evidence of in the current discussion of the subject. Every institution has a life of its own, and different institutions have different requirements of health and well-being; they cannot all be expected to meet the same kind of requirement in recruiting their personnel. The characteristic Indian view of democracy would seem to require that every public institution should acquire something of the character of a multi-caste panchayat. Our political parties might survive—or even gain from—the experiment to achieve parity between castes: it is doubtful that our scientific research laboratories can.

Separate from the quantum of reservation is the question of its duration. It cannot be too strongly emphasized that the makers of our Constitution thought of reservation as a temporary measure. No doubt the makers of the Constitution were optimistic in thinking that reservation could achieve its objective in ten years; but the basic question is whether a process, which they clearly saw as reversible, has now got to be accepted as irreversible. If we are to continue to regard reservation as a means to an end, the end must be reconsidered and redefined; or it may cease to be a means to an end, and become an end in itself. The metaphor of 'crutches' has been applied to reservation; certainly, in South India the Other Backward Classes have on occasion used the crutches as sticks to beat the system with.

Ours is a divided society, with extremes of inequality between castes and communities inherited from the past. It does not take much imagination to see that job reservation can do little to bring about a social revolution in India. Even its strongest advocates do not believe that job reservation has achieved very much for the common man during the last thirty years, or that it will bring about any miracle during the next thirty years. But many people seem to think that if job reservation benefits only a few, it is at very little cost to the rest. This is a mistaken view. Reservation in the right sphere of public life may do a little good; in the wrong sphere of public life it may do a lot of harm.

It is necessary to underline the implications of the distinction between reservation in positions that are filled by election and reservation in positions that are filled by appointment. Our Constitution makes that distinction clearly. It speaks of the reservation of seats for the Scheduled Castes and the Scheduled Tribes in Parliament and in the Assemblies, but only of their claims to 'services and posts in connection with the affairs of the Union or of a State'

It lays down the proportion of seats in the elected bodies that should be reserved for members of the Scheduled Castes and the Scheduled Tribes, but avoids presenting quotas or even reservations in the services. The constitutional provisions relating to the place of the Scheduled Castes and the Scheduled Tribes in the legislatures, being specific, are also time bound; the provisions relating to their place in the services are neither specific nor time bound.

Those who are elected to Parliament and to the Assemblies are expected to perform a representative function. They represent a particular constituency, and it is their obligation to speak up for that constituency, and to protect and promote its particular interests. The obligations of the civil servant are different. The civil servant does not perform a representative function in the way in which members of elected bodies do. To expect the civil servant to protect or promote the particular interests of his caste or community is to set him at odds with the basic requirement of his service. A legislature can to some extent legitimately become an arena for the adjustment of interests between castes and communities; a civil service cannot.

Politicians who are elected to a representative body from a reserved constituency owe a special obligation to the castes and communities in whose name that constituency is reserved; to ignore the special claims of these castes or communities would be a breach of faith on their part. But is it defensible for the civil servant who is appointed on a community quota to give special attention to the interests of his community in the manner of an elected representative? If not, then the argument that the weaker sections of society need special representation in the civil service on a caste or community basis becomes somewhat weakened.

Administration may be conducted through a variety of arrangements, and we have consciously adopted a bureaucratic, as opposed to a feudal or a patrimonial, system of administration. A bureaucracy is governed by a set of norms for regulating both its internal life and its relations with the outside world. These norms call for dispassionate conduct in accordance with impersonal rules. A good official is one who deals with his superiors and subordinates according to the rules of his office, without attention to personal considerations; a good civil servant is one who serves all members of the public alike, without regard for the claims of caste and community. It is well known that the civil servant in India is under constant pressure to show special favour to his kinsmen, his caste-folk and his co-religionists; but we commend him not for yielding to such pressure, but for rising above it.

Thus, in reconsidering reservation, we must consider the effect of reservation not only on the numerous castes and communities which constitute the backward classes, but also on the civil service as an institution. We must begin with an assessment of those states in South India which have quotas in government appointments not only for the Scheduled Castes and the Scheduled Tribes, but also for a whole host of other castes and communities grouped together as the Other Backward Classes. Where half the appointments to posts in the state government have to be made in accordance with a roster of castes and communities, this cannot leave unaffected the nature of the service itself. It would be idle to talk simply in terms of efficiency, for what is in question is the very character of an impartial civil service designed to serve the public, irrespective of caste and community.

If the obligations of office are different for the civil service as compared with the legislature, the professions—such as law, medicine and teaching—are in turn bound by their own distinctive norms and values. It is true that university professors, like civil servants, are appointed—and through selection procedures that are analogous to those used in the recruitment of civil servants. It is a sign of the weakness of the academic profession when university professors are recruited, as they are in some states, through the state public service commission, and when their service conditions are modelled on the civil service. This tendency has become greatly strengthened in Kerala, Karnataka and Tamilnadu by the universities being required to fulfil caste quotas in their appointments in conformity with the pattern established by the state public service commission.

Democracy cannot prosper in the absence of a certain respect for the autonomy of institutions. This autonomy is subverted when a central agency—whether legislative or executive—decides that the internal life of every institution will be regulated by a single plan, irrespective of the nature and functions of the institution concerned. If civil servants are expected to behave increasingly like politicians, professionals are expected to conduct themselves like government employees. It would be too much to attribute all of this to the policy of reservation, but it is in this light that we have to see the recent threat of the U. G. C. to cut off funds to the universities unless they toe the official line on reservations. In almost every important matter, civil servants try to browbeat professionals in exactly the same way in which they are browbeaten by politicians.

The academic profession differs from the civil service in at least two important regards: the rewards due to merit and to seniority are balanced differently in the two cases, and so also are the demands of personal as opposed to impersonal relationships. An academic institution would lose much of its credibility if it made its promotions by seniority alone, leaving it to God or to posterity to reward merit. The quota system sets beside the principle of merit, some other principle in the name of social justice. But the more persistently the claims of academic merit are overlooked within the university, the more steadily people lose faith in the very things they come to the university to do.

Whereas the relationship between the bureaucracy and the public it serves is or ought to be impersonal, there is a large personal component in the relationship between teachers and students. Any healthy university must keep some room for personal attachments between students and teachers, and for personal rivalries among teachers and among students. No great harm is done where these centre around academic issues. But what would a university be like if attachments and rivalries were to follow the lines of caste and community? If the quota system is designed to ensure that students belonging to particular castes and communities are not discriminated against by teachers belonging to other castes and communities, will it be possible to prevent the university from breaking up into warring factions protecting and promoting the interests of community and caste? It would be naive to believe that caste and community can be allowed to stand between teacher and student, between doctor and patient, or between scientist and technician without transforming the very character of the university, the hospital or the laboratory.

What has gone wrong with our thinking on the backward classes is that we have allowed the problem to be reduced largely to that of job reservation. The problems of the backward classes are too varied, too large and too acute to be solved by job reservation alone. The point is not that job reservation has contributed so little to the solution of these problems but, rather, that it has diverted attention from the masses of Harijans and Adivasis who are too poor and too lowly even to be candidates for the jobs that are reserved in their names. Job reservation can attend only to the problems of middle class Harijans and Adivasis: the overwhelming majority of Adivasis and Harijans, like the majority of the Indian people, are outside this class and will remain outside it for the next several generations.

Today, job reservation is less a way of solving age-old problems than one of buying peace for the moment. It would be foolish to blame only the government for wanting to buy peace in a country in which everyone wants to buy peace. It would be foolish also to recommend an intransigent attitude to a government which has neither the will to impose its power nor the imagination to think of alternatives. But unless it is able to offer something better to the backward classes than it has done so far, reservation will continue to bedevil it. It is for this reason that we need a discussion of the whole range of disabilities from which the backward classes—and in particular the poorest and the lowliest among them—suffer.

The point has to be made that those who are opposed to job reservation are not for that reason against the recognition of the special claims of the Scheduled Castes and the Scheduled Tribes. Where there are special claims, special provisions have to be devised for meeting them. It is here that one must distinguish as clearly as possible between those special provisions which benefit directly only a few individual Harijans and Adivasis and those which benefit Adivasis and Harijans as a whole—or, at least, large numbers of them.

Adivasis and Harijans suffer above all from the burden of poverty and the stigma of pollution. There is no simple cure for either of these two evils. But a little more can be done than has so far been done for relieving the destitution of the millions of Harijan and Adivasi labourers, particularly in the rural areas, whose conditions of life and work have been unaffected—if not affected adversely—by the course of economic development. The welfare of the backward classes can be made the starting point of economic planning instead of being a minor branch of it as it has been so far. Reserving a few more jobs in the government for the backward classes and making plans for the economic advancement of those who are already advanced are two sides of the same coin; the basic prerequisite for an improvement in the material condition of the backward classes is a change in our approach to planning as a whole.

Millions of Harijans and Adivasis still labour under semi-servile conditions in the rural areas. It is true that social prejudice acts against the backward classes in the urban areas as well—in the neighbourhood and in the office; but this is mild in comparison with the brutality to which Adivasis and Harijans—men, women and children—have to submit in the village and on the farm. A great deal remains to be done—and something can be done at moderate cost—to change the conditions of work in those occupations such as scavenging, flaying and tanning, which continue to be tainted by the stigma of pollution and in which many Harijans are still engaged. From time immemorial the stigma of pollution has been associated with degrading conditions of work, and the stigma will remain so long as the conditions of work are unchanged.

The stigma of pollution is not the only factor behind the degradation of the backward classes; the absence of literacy and education is also important. There can be no significant change in the conditions of the Scheduled Castes and Scheduled Tribes unless they acquire a measure of self-reliance, and they cannot as a collectivity acquire this self-reliance if most of their individual members remain ignorant and illiterate. If special efforts are to be directed towards the backward classes, then these must concentrate on the creation of conditions for literacy and primary education out of which the largest number of individual Harijans and Adivasis can benefit. There are changes that can be brought about within a generation in the conditions of millions of people, and it is these that should receive the first priority. It is no doubt cheaper to provide the best of higher education to a few selected members of the backward classes, but we must look not merely to the costs but also to the benefits of spreading literacy and primary education widely, if not universally, among the Harijans and the Adivasis.

The plain fact is that we cannot spread literacy, education and decent conditions of work among the Harijans and Adivasis without also spreading them among the weakest sections of Indian society without regard for tribe, caste and community. In the end, the backward classes will gain and not lose if the benefits due to the disprivileged are spread widely instead of being confined narrowly to a few selected individuals. Moreover, there can be no real change in their conditions without a change in the structure of Indian society and a corresponding change in the mental horizons of people. We will remain eternally stuck with the very divisions we have inherited from the past unless we are able to move towards a different conception of backwardness, one that relates it to the needs of individuals rather than the demands of castes and communities.

II

The Indian Road to Equality: More Jobs for More Castes

The unanimous and enthusiastic endorsement by Parliament on 11 August of the Report of the Backward Classes Commission constitutes an important landmark in the history of contemporary India. If Parliament has acted in full awareness of the likely consequences of its action, we are perhaps entering a new phase in the reconstitution of Indian society. This reconstitution may be no less far-reaching in its scope than the one attempted by the new Constitution which Indians fashioned for themselves on achieving independence.

What was acclaimed by all parties in Parliament was the spirit of equality embodied in the Commission's Report. The Constitution too is marked by its emphasis on equality, and in that sense our present Parliament can claim to be carrying forward the task begun by the Constituent Assembly. But there the similarily ends. The

The Times of India, 28 August 1982.

aim of the Constituent Assembly was to bring about a casteless and a classless society. The aim of the Commission under Mr B. P. Mandal was, presumably, to exorcize the disparities of class by placing caste at the centre of Indian society.

It will be a mistake to miss the moral impulse behind the acclaim with which the Report was greeted in Parliament. Indians are too ready to mistrust their politicians and to attribute the worst motives to their every action. It is not true that support for job reservation, which is the principal recommendation of the Commission, is motivated solely by the calculation to gain something for oneself or one's nephews or one's own community. Job reservation is coming to be viewed sympathetically by more and more people who have nothing to gain personally from it. The moral impulse behind it is the commitment to equality. Indians today seem to be ready to do anything in the cause of equality, even to rehabilitate the caste system, and this is the case with all political parties and with intellectuals as well as politicians.

Equality and distributive justice are universal concerns in the modern world, and it has been said that these concerns were activated among Indians by their exposure to the world outside, particularly the western world. The Indian Constitution has relied heavily on other constitutions, particularly in the various guarantees of equality written into it. What seems to be unique therefore is not the Indian concern for equality, but the means being increasingly recommended for the attainment of equality in India. The rehabilitation of caste may not be the intention of those who recommend job reservation, but it will certainly be a consequence of their recommendations if these are seriously acted upon.

Mr Mandal's Commission was not the first to be set up in independent India for recommending measures for the betterment of backward classes *other than* the Scheduled Castes and Scheduled Tribes. The first Backward Classes Commission was set up in 1953 under Kaka Kalelkar and it submitted its Report in 1955. Nothing very much came of that Report which was equivocal in its attitude to caste, unlike the present Report which is clear in recommending caste as the basic unit in all considerations of distributive justice. It is not surprising that some members of Parliament wanted that, in addition to the castes and communities listed in the Report, religious minorities should also be brought within the scope of its recommendations.

In the Constituent Assembly Dr Ambedkar had argued for adopting the *individual* as the unit in the new scheme of things: compensatory measures for the Scheduled Castes and Scheduled Tribes were regarded as *exceptional* and *transitory*. At about the same time, Dr Radhakrishnan's Commission on University Education had asserted: 'The fundamental right is the right of the individual, not of the community'. In boldly asserting the claims of caste, Mr Mandal's Commission has brought back an alternative conception of Indian society whose appeal may go far deeper than the appeal of the conception on which the Constitution is based. Indian intellectuals do not like talking about caste, but that does not mean that they are not moved by its appeal.

The first Backward Classes Commission had come too soon after the Constituent Assembly to be able to advocate the cause of caste and community without any reserve. The Commission as a whole was in favour of extensive reservations on the basis of caste. But its Chairman, Kaka Kalelkar, could not in the end bring himself to endorse a recommendation which he felt would be bound to lead to a rehabilitation of caste. In his covering letter to the President, he pointed out that in a democracy official recognition could be given to only the individual and the nation, and added that he was 'definitely against reservation in Government Services for any community'.

The recommendation of the first Backward Classes Commission was thus infructuous, but reservation in the services on the basis of caste and community has continued to flourish, particularly in the states of South India. The Supreme Court, after resisting caste quotas for some time, has taken a more latitudinarian stance in recent years. Various commissions have been set up in the states to recommend measures for the betterment of the weaker sections, and these have almost invariably recommended the reservation of more jobs for more castes. The Karnataka Backward Classes Commission under Mr L. G. Havanur has sought to provide the most comprehensive rationale for caste quotas in every sector of employment, arguing that, far from being unconstitutional, the official recognition of caste is in fact required by the Constitution itself.

To the makers of the Constitution equality meant, above all, equality of opportunity which takes only individual merit into account without any consideration for caste or community. As against this is the idea of the equal distribution of all benefits, particularly the benefits of education and employment, among the castes and communities into which society is divided. The two conceptions are different, and in their application mutually incompatible. It is difficult to predict how far equality between castes will in fact be achieved by the kind of redistribution recommended by the Commission. But any attempt at a redistribution which seeks to satisfy the claims of castes is bound to strengthen rather than weaken the hold of caste over public life.

Why have Indians become more sympathetic to the claims of caste between the Commissions of Kaka Kalelkar and Mr Mandal? Firstly, people are left with fewer illusions today than they probably had thirty years ago as to how far equality of opportunity can carry them in this society. Politicians, civil servants, engineers, doctors, lawyers, professors, journalists have all come to believe that, whatever the case may be elsewhere, in India competition is neither free nor fair. If you tell them that reservation penalizes merit and ability, they will tell you that merit and ability are in any case penalized in this society. Reservation tries at least to make some reparation for the injuries suffered by disprivileged groups in the past.

There is also the argument that equality is a universal value, but that we must find an Indian way to make it work in this society: Indians have become increasingly conscious of the need to find 'alternatives' in every sphere. If very little progress has been made so far in the attainment of equality, this may be because the methods adopted, being primarily western in inspiration, are not well suited to the Indian social environment. For better or for worse, this environment has always favoured the community over the individual, and it may be possible in India to combine the preference for the community with the commitment to equality. The west may go its own way, but we in India will have *both* equality *and* caste.

It is said that extensive reservation will lead to a decline in the efficiency of our public institutions. There are other, more important, consequences to be considered. Reservation on the scale being proposed will alter the character and not merely reduce the efficiency of these institutions. The object of reservation is to provide equitable representation in them to all castes and communities. But not every institution performs representative functions or is designed to perform them. A bank or a hospital or a research laboratory has a very different job to do from a political party or a district council. It may be legitimate for a member to represent the interests of his caste or community in a legislative assembly. Can we legitimize the representation of caste interests in institutions such as the universities without changing their character altogether? The maintenance of efficiency is in my judgement less important than the protection of the very norms by which alone such institutions can be governed.

III

Caste and Politics: The Subversion of Public Institutions

In assessing any scheme of reservations today, we have to keep in mind the distinction between those schemes that are directed towards advancing social and economic equality, and those that are directed towards maintaining a balance of power. Reservations for the Scheduled Castes and Scheduled Tribes are, for all their limitations, directed basically towards the goal of greater equality overall. Reservations for the Other Backward Classes and for religious minorities, whatever advantages they may have, are directed basically towards a balance of power. The former are in tune with the spirit of the Constitution; the latter must lead sooner or later to what Justice Gajendragadkar had called a 'fraud on the Constitution'.

Supporters of the Mandal Commission's recommendations would

The Times of India, 11 September 1990.

like to group the Other Backward Classes together with the Scheduled Castes and Scheduled Tribes. This is an error of judgement, not always made in innocence. The Other Backward Classes have a very different position in Indian society from that of the Scheduled Castes and Scheduled Tribes. It is true that their traditional ritual status was low and that they were latecomers to the competition for university degrees and government jobs. But only the Harijans and Adivasis have been for centuries the victims of active social discrimination, through segregation in the first case and isolation in the second. They alone have suffered the kind of psychological and moral injury that justifies their being now treated with special consideration. The castes and communities grouped together as the Other Backward Classes have not suffered collectively that kind of injury in either the recent or the distant past. They include locally dominant castes, some of whose leaders are among the worst tormentors of Harijans in the rural areas today.

The Other Backward Classes should be grouped, not with the Scheduled Castes and Scheduled Tribes, but with the religious minorities. The students' union of Aligarh Muslim University drew attention to this obvious fact when it welcomed the recommendations of the Mandal Commission and added that the benefits proposed by it should be extended also to the Muslims. If the idea behind reservations is that power should be so distributed as to maintain a balance between all castes and communities, then the plea of the Muslim students is not unreasonable. But that was not the basis on which reservations were adopted by the Constitution; the basis there was to provide special opportunities for the most disadvantaged sections of society.

It is essential to remember that reservations were first introduced by the British. The colonial policy of reservations, which was administered extensively in peninsular India, was governed mainly by considerations of the balance of power. That being the case, there was nothing incongruous about the fact that the old Madras Presidency had 100 per cent reservations for a time, or quotas for both 'forwards' and 'backwards'. There was indeed a certain unity and coherence in the policy of the colonial administration: job quotas for all on one side, and separate electorates for religious groups on the other.

The colonial administration was not in any way hampered by the principle of equality of opportunity. Article 16 of the Constitution (together with Article 15) altered dramatically the scope for reservations in independent India. What was possible under the Brtitish was no longer possible in the new order. Reservations could now be justified only by the argument for greater equality and not by any argument for the balance of power. The provisions in the Constitution relating to job reservations do not come anywhere near to even mentioning numerical quotas. Those on reservations in the legislatures specify such quotas as clearly as possible, but are, on the other hand, strictly time-bound. The Constitution maintains a consistent distinction between the two kinds of reservations, dealing with them in two separate places. But the executive, with some help from the judiciary, has succeeded in obscuring that distinction in the public mind.

The Indian Constitution is committed to two different principles both of which relate to equality: the principle of equal opportunities and the principle of redress. It is difficult, even under the best of circumstances, to evolve a coherent policy that will maintain a satisfactory balance between the two. The present policy of massive numerical quotas in public employment is a perverse application of the principle of redress that threatens to eat up the principle of equality of opportunity.

The principle of redress is a broad one that can be translated into many kinds of affirmative policy action. For instance, there can be substantial investment, on a preferential basis, to raise the levels of health, housing and elementary education among the weaker sections of society. Job quotas in public employment are by no means the best way of reducing social and economic disparities between castes and communities, and they have serious institutional costs that far outweigh the benefits they bring to some individual members of backward castes.

Preferential policies may be considered as a way of creating special opportunities for some over and above the equal opportunities available in principle to all. The tension between 'special opportunities for some' and 'equal opportunities for all' is too obvious to be ignored. If that tension is to be kept under control and not allowed to subvert the institutional system, preferential policies must be used judiciously and with restraint.

Paradoxically, caste has increased its hold over public life, despite such modernization as there has been in India since independence. At the time of independence, many Indians believed that caste was on its way out, and they had some evidence to support their belief. The many ritual rules by which distances between castes had been maintained in the past were declining or dying out. The restrictions on commensality were rapidly breaking down; marriages were taking place between subcastes of the same caste, and it was hoped that this would show the way to intercaste marriages on a wider scale. However, there was one sphere of life, that of politics, in which caste not only held its ground but began to strengthen its hold. If caste has acquired a new lease of life in independent India, this is almost entirely because of the increasing use made of it in politics.

In the last forty years, and particularly since 1977, a tacit consensus seems to have emerged that all political bodies—*zilla parishads*, state cabinets, party committees—should be so constituted as to represent the major castes and communities. Representation in India has come increasingly to mean the representation of castes and communities. Two questions arise from this. The first relates to the purely political domain: does a political body become representative only when its composition matches the distribution of castes and communities in the larger society? Secondly, should all public institutions, irrespective of their functions, be constituted in the manner of political bodies? What is disquieting is the growing belief among leaders of all political parties that if caste balances are good for the domain in which they operate, they ought to be good for all institutional domains.

Much of the current debate on reservations is focussed on the question of employment: how much employment can and should be provided to our educated men and women, whether new jobs can be created to maintain the existing level of employment among the forward castes despite job reservations, and so on. But there are other issues, besides the distribution of jobs among castes, that we must consider if we are to assess the long-term implications of reservations. The most serious implication of extending caste quotas is that all public institutions will by that process come to be cast in the mould of political councils and committees. Universities, hospitals, scientific research laboratories, defence establishments and courts of justice will all come to look more and more like representative political bodies. If, like our present political bodies, they are constituted so as to maintain a balanced representation of castes, it will not be long before they begin to function like them.

Every public institution will then be riddled not only with caste but also with politics, for caste has no place in these institutions except as an instrument of politics.

IV

Caste and Class: Some Avoidable Misconceptions

There has been much talk recently of the relative importance of caste and of class, particularly in the context of the strategies to be used in identifying the most disadvantaged members of society. Undoubtedly, there is a relationship *empirically* between the two in Indian society, but caste and class are *analytically* distinct, and the distinction has to be clearly understood. Caste is in many ways a specifically Indian phenomenon, and, although we know much about it from experience, it is not easy to define it in such a way as to enable us to distinguish it clearly from other forms of inequality or stratification. Class, on the other hand, is a more general social phenomenon whose nature and forms have been extensively studied in many parts of the world. Some say that caste is the form taken in

The Hindustan Times, 4 October 1990.

Indian society by class; but this is a vague formulation that can lead to serious mistakes in the representation of social processes.

Since it is an aspect of many societies, class has been a subject of extended discussion in social theory. This does not mean that social theorists have reached complete agreement about the meaning to be assigned to the concept. I will first indicate a number of differences before I turn to the commonalities. There is, first, the difference in approach between Marxists and bourgeois social theorists. The former take property, or position in the system of production as their starting point, whereas the latter start with occupation, education and income; clearly, there is much scope for combining these different criteria in formulating a definition of class.

Criteria such as wealth, income, occupation and education are extremely important in differentiating not only one individual (or household) from another, but also one subdivision of society from another. We can use such 'socio-economic' criteria, singly or in combination, to differentiate and rank language groups, religious communities, castes, sects, tribes, and so on. That would no doubt reveal a great deal about the society in question, but it would not provide an account of the social classes in it. No matter how carefully we select our socio-economic criteria and how diligently we apply them, a ranked set of language groups, denominations, sects, castes or tribes is in principle different from a set of social classes, if we go by the meaning assigned to the term 'class' in social theory.

Thus, in comparing caste with class, we have to keep two things in mind: the criteria that we apply to differentiate and rank the component parts of a population, and the units to which we apply those criteria. The basic and irreducible units with which classes are constructed are individuals (or at most households). A social class is a set of persons whose individual members (either in their own right or as heads of households) are assigned to it because they fall within a certain range of variation. We get a set of classes only when we start with individuals (or households) and place them together on the basis of one or more economic or other social criteria. The grouping of castes, or tribes, or races in terms of even purely economic criteria does not give us classes, but ranked divisions of some other kind. It is wholly inconsistent with sociological usage to take groups of castes, tribes and communities and to call them 'classes' on the ground that the groupings have been made on the basis of 'economic' criteria. The crucial point here is not what kind of criteria we apply, but what kind of units—whether individuals or groups—we seek to differentiate and rank by the application of those criteria.

Now in jurisprudence, the term 'class' has a much wider meaning than in social theory, and it would be absurd for the sociologist to insist that his concept should prevail over the lawyer's. That term is in fact given the widest possible meaning in logic and in mathematics: in logic, a class is simply the outcome of any consistent scheme of classification. We may thus take any universe of objects, such as persons, animals, plants, rocks, or even numbers, and classify its members by the consistent application of one or more criteria; we will then have classes of numbers, rocks, plants, and so on. But most of these will have very little to do with what are designated as classes in social theory. The logical concept of class is wide and general; the sociological concept, restricted and specific.

Human beings may be classified in all sorts of ways with the use of economic, psychological, biological and other criteria. Physical anthropologists use criteria such as stature, head shape, hair form, etc in their classifications, so that we can speak of a class of short-statured, medium-headed persons, a class of medium-statured, long-headed persons, and so on. Likewise, psychologists may classify the population in terms of IQ, learning ability, emotional stability, and so on. These are all classes in the general sense that they are the outcome of classification, but not in the specific sense in which the social theorist, whether Marxist or non-Marxist, uses the term 'class' to compare and contrast it with caste.

Now, it should be possible to classify in this way not only individuals, but also groups of individuals: we will then have classes of such groups, the latter themselves being defined by race, caste or ethnicity. An important condition for a sound classification is that the groups classified should be internally homogeneous. Where the groups themselves lack internal homogeneity, the classification will be weak and defective. I may illustrate the point quickly by taking the example of the classification of races. Although physical anthropologists have come out with many schemes of classification, it has turned out in almost every case that variations within the class are larger than those between classes, leading some of them to abandon the very prospect of arriving at a satisfactory classification of races.

There should be no objection in principle to classifying groups of castes according to one or more social or economic criteria. If the classification is sound, each such group may be called a class, but we must remember that we will then be using the term in its broad, logical sense and not in its restricted sociological sense. According to the latter, a social class is composed of individuals each of whom is at a certain socio-economic level, and not of groups whose average socio-economic level falls within a certain range. If we make a classification of castes in terms of the average socio-economic levels of their individual members, we will naturally find that there are disparities between these averages, such that some groups of castes have a high average whereas others have a low one. But even in castes with a low average level, there will be some, if not many, individuals (or households) with a high or very high level; and vice versa. The point simply is that we get a different kind of classification when we start with individuals or households from what we get by starting with castes or communities.

A classification of castes in terms of the average socio-economic standing of its individual members cannot be regarded as unsound in principle. Much will depend on the extent to which the 'classes' so constructed are internally homogeneous. From the logical point of view, a classification is suspect if variations within the class are larger than those between classes. I have already alluded to the difficulties encountered by physical anthropologists in their attempts to classify races in terms of biological criteria. Attempts to classify castes in terms of socio-economic criteria are likely today to encounter similar, if not larger, difficulties.

From the viewpoint of jurisprudence, a classification may be logically consistent and yet fail to meet the requirement of a reasonable classification. What, then, is a reasonable classification? A reasonable classification I take to be one in which there is a clear nexus between the scheme of classification and the object for which it is made. It is possible to classify castes in many different ways, depending on the object. The classification of castes as 'Depressed Classes', 'Backward Castes', and so on was introduced during British rule. New terms or categories, such as 'Scheduled Castes' and 'Most Backward Castes' were adopted subsequently; and the distinction between 'backward' and 'forward' castes has gained currency in many parts of the country. The policy objectives behind such classifications may be said to be to 'redress the bias of contingencies in the direction of greater equality'. Can they be expected to meet that objective efficiently.

Where job reservation is made the primary instrument of a policy of redress, two things must be kept clearly in view. Firstly, the number of reserved jobs is very small in relation to the population made exclusively eligible for them. Secondly, there is a very high probability that the reserved places will be taken not by the average members of the backward castes who may be poor or illiterate or live in thatched huts, but by the best-off members of those castes who are more than likely to be of a different social class from their poor brethren. The general correspondence between caste and class is beside the point: it is precisely where that correspondence breaks down that the gains of reservation are likely to flow. It is there that the nexus between the classification of castes and the object of the classification has to be submitted to the closest possible scrutiny.

V

The Gains of Reservation: A Comparative Perspective

One of the main objectives the country set before itself at the time of independence was to reduce if not remove inequality through a process of planned social change. Inequality was conceived in a broad sense to include not only the disparities of income and wealth but also the traditional feudal or hierarchical relations between people and the attitudes sustaining those relations. Progress in realizing the objective of greater equality has been halting and uneven, partly because of a lack of will and partly because of obstacles whose nature and significance had not been foreseen in advance.

Among the many efforts to bring about a measure of equality in a society saturated with hierarchical distinctions, the programme of positive or compensatory discrimination for the backward classes

The Times of India, 6 September 1984.

occupies an important place. In its scope and reach it stands comparison with the programme of agrarian reform. Positive discrimination, like agrarian reform, has received considerable attention from the legislatures, the courts and the executive, and both have given rise to a voluminous literature. In what follows I shall discuss the problem of assessing the part played by positive discrimination in reducing social disparities.

Positive discrimination may be described as a way of reducing social disparities by creating special opportunities for some in addition to the equal opportunities created for all. Obviously there is a tension between the two principles—equal opportunities for all and special opportunities for some—and success in realizing the larger objective of reducing disparities will depend on the care with which a balance is struck between the two. It is thus that one may support the creation of special opportunities in certain spheres for the Scheduled Castes and Scheduled Tribes while at the same time opposing the creation of such opportunities in the same spheres for the Other Backward Classes.

Positive discrimination in favour of the Scheduled Castes and Scheduled Tribes has generated less public controversy in independent India than positive discrimination in favour of the Other Backward Classes, and I shall concern myself solely with the latter. These Other Backward Classes are not all defined in exactly the same way in the different parts of the country, but by and large they are an assemblage of castes or caste-like groups. The special measures adopted or recommended for their advancement are many and diverse, but by far the most important among them are reservations in educational institutions, especially in professional and technical colleges, and in government service.

Opinion differs quite widely regarding the social benefits as well as the social costs of reservations for the Other Backward Classes. In the years immediately following independence, opinion in the country, at least outside peninsular India, was on the whole unfavourable to the extension of the benefits of reservation to others besides the Scheduled Castes and Scheduled Tribes. The first Backward Classes Commission recommended extensive reservations in 1955, but there were several notes of dissent and the Chairman of the Commission himself strongly attacked the reservation of posts in government on the basis of caste. Throughout the fifties and into the sixties the union Home Ministry opposed the use of caste in making concessions to the Other Backward Classes.

The tide began to turn some time in the seventies. In several North Indian states leaders of intermediate castes began to claim special concessions for their members on the ground that they were being excluded from the benefits of development. The crucial position occupied by Mr Charan Singh in the Janata government helped to give these claims a focus. A new Backward Classes Commission was set up by the union government in January 1979 under the chairmanship of Mr B. P. Mandal which submitted its Report in December 1980. The second Backward Classes Commission has shown none of the hesitations and equivocations of the first. Mr Mandal has recommended extensive and comprehensive reservations in favour of a large number of castes and communities throughout the country as a necessary if not inescapable step towards the realization of social equality.

Earlier recommendations regarding the Other Backward Classes, whether for or against reservations, were frequently criticized for lacking an adequate empirical basis. Mr Mandal's Commission has sought to make good this lack by collecting and compiling a large body of data on a variety of topics. It has tried to use these data to forestall all possible objections to the use of caste quotas in education and employment. More recently, an American student of the subject, Marc Galanter has brought together a formidable body of material in a book which might give some support to Mr Mandal's case. Professor Galanter seems to follow Mr Mandal in his view that objections to reservation as a method of attaining equality arise largely from lack of information and from bias among the intelligentsia who belong largely to the upper castes.

For all the work put in by his Commission, Mr Mandal has ignored certain obvious questions which might in effect undermine his advocacy of caste quotas as a necessary step towards equality. If the use of caste quotas is the only way of reducing social disparities between groups, what has been the record of those states which have refused to use such quotas except for the Scheduled Castes and Scheduled Tribes? There are such states in India, and it cannot simply be assumed that they are all indifferent to the objective of reducing social disparities.

West Bengal offers a test case to those who maintain that there is no way of reducing social disparities except through positive discrimination, and it is remarkable that both Mr Mandal and Professor Galanter have so little to say about it. Successive governments in that state have refused to introduce reservations for the Other Backward Classes, and it does not appear that the people as a whole in West Bengal have been greatly aggrieved by this. A Committee was set up on 1 August 1980 to look into the problems of the Other Backward Classes and it submitted its report within a month, recommending against the use of caste quotas in government service in favour of the Other Backward Classes.

Mr Mandal's Commission had a comparative study made by the Tata Institute of Social Sciences of the four states of Karnataka, Tamil Nadu, Bihar and Uttar Pradesh. Since these states have all adopted caste quotas in one form or another, the study failed to consider whether there are viable alternatives to reservation for reducing disparities in society as a whole. A comparison between, say, Karnataka and West Bengal would have revealed not only what reservations have actually achieved but also what may be achieved without reservations.

There are no doubt enormous difficulties in comparing regional patterns of inequality in a systematic and objective way. But on the face of it there is little reason to believe either that West Bengal has been less concerned about reducing social disparities or that Karnataka has actually achieved greater success in that respect. Here we have to take into account not only disparities in the distribution of wealth and income, but all those inequalities in social life which were characteristic of the traditional hierarchical order. There is enough evidence of significant changes in social relations and social attitudes in West Bengal as in other states like the Punjab which have shown little enthusiasm for job reservation.

If we look at regional variations in patterns of positive discrimination it will be obvious that the decision to adopt reservations on a large scale is governed by various considerations, although an attempt is invariably made to justify the decision by an appeal to the principle of equality. Political parties have not adhered to a uniform position on this. In West Bengal neither the Congress nor the Communist party has shown much enthusiasm for reservations, whereas in Kerala both the CPI and the CPI(M) have supported caste quotas as has the Congress in Karnataka.

Variations in patterns of reservation tell us less about regional structures of inequality than about regional political alignments. Reservations were introduced into peninsular India during British rule less as an instrument for attaining equality than as a device for maintaining a balance of power between the various communities. It is now forgotten that what the Supreme Court struck down in the Champakam Dorairajan case in 1951 was not so much a rule of positive discrimination as a system of communal quotas. Under the old dispensation in Madras *all* positions in specified areas were filled by rotation between the different communities. This is no longer possible under the new Constitution. But where, as in Karnataka, more than two-thirds of the positions are reserved under the umbrella of positive discrimination, one is only a short step away from a system of communal quotas.

VI

Resistance to Reservations: Some North-South Differences

The intelligentsia in India stands divided between those who support and those who oppose the recommendations of the Mandal Commission for reservations in public employment. In Delhi and other north Indian cities, the majority of professors, journalists, lawyers and doctors are today opposed to the further extension of caste quotas in favour of the Other Backward Classes; but that does not necessarily mean that they are going to have the last word.

There have been arguments in support of caste quotas from left intellectuals as well as some who claim allegiance to the ideals of Mahatma Gandhi. But attempts to justify caste quotas in terms of Marxian or Gandhian principles appear more surprising than convincing. If we are to defend the recommendations of the Mandal

The Hindu, 20 October 1990.

Commission, we must frankly admit the limitations of those principles, and look for our arguments elsewhere. The most persuasive argument that I have heard in support of extensive caste quotas in public employment is a functional one, namely, that they have worked successfully for decades in south India, and, therefore, they ought to work successfully in India as a whole. That argument deserves some serious consideration from those who are opposed to the recommendations of the Mandal Commission on grounds of principle.

It is certainly the case that caste quotas in public employment have prevailed in south India more or less continuously and in some parts very extensively for nearly seventy years. But to say that something has prevailed for a long time is not to prove that it has worked successfully. Even the most ardent advocates of caste quotas cannot argue that they have worked equally successfully in all areas of public life in the south. If we are to reach a balanced assessment of the consequences of job reservations, we have to adopt a differentiated approach and consider separately and in turn their operation in institutions of different kinds.

Let me begin with institutions devoted to science and scholarship, since those are the ones closest to my experience as an academic. Here I would include both universities and institutes of advanced study and research. South India has a number of institutions devoted to science and scholarship whose work enjoys a high reputation throughout the country and abroad: the Indian Institute of Science in Bangalore, the Centre for Chemical and Molecular Biology in Hyderabad, the Centre for Development Studies in Trivandrum and the Madras Institute of Development Studies. These are precisely the ones that are substantially, if not fully, insulated from caste quotas in faculty appointments. Compare with them the south Indian universities in which faculty appointments have to be made in accordance with a roster system. No doubt caste quotas prevail in south Indian universities; but can they be said to work successfully? I would certainly reconsider my judgement on reservations if I were to find even a few institutions in which sustained academic work of a high order is being done despite the extensive use of caste quotas in faculty appointments.

It is difficult to make comparisons of that kind for the services of the state government, because reservations prevail everywhere so that we cannot say whether government departments in a particular south Indian state work better with or without caste quotas. We can, of course, compare general administration in a south Indian state with its counterpart in a north Indian state, and the comparison will in many cases, though not in all, be probably to the advantage of the former. But we cannot from that conclude directly that, say, Tamilnadu is better administered than Bihar because caste quotas have been in force there for a longer time. All that we can say perhaps is that there are many ways to ruin administration, and not just through caste quotas. There is another point that must not be lost to sight. The IAS and other central services provide the apex of administration in all states in the country, in the north as well as the south, and these are still free from caste quotas with the exception of those in favour of the Scheduled Castes and Scheduled Tribes. We cannot make a final judgement on the efficacy of administration in south India without taking into account those cadres to which recruitment is still largely made through open competition.

But after all that has been said, the fact still remains that extensive caste quotas in favour of the Other Backward Classes have been in force in the south for a long time, and are widely accepted, or at least tolerated there, whereas the prospect of their introduction has aroused unprecedented resentment and hostility in the north. How are we to account for this difference in attitude and orientation among people in two different parts of the same country? Here I can only indicate a few lines of enquiry that may lead to an answer to this important question.

How well a given arrangement works, or whether it works at all, depends not only on its intrinsic merit but also on the circumstances and the manner of its institution. The circumstances under which caste quotas were imposed in south India in the high noon of colonial rule were totally different from the ones under which they are being sought to be instituted today after more than forty years of national independence. It will be a serious mistake to assume that nothing has changed in the outlook and orientation of people in the last seventy years. There have been changes not only in their attitudes to caste, but, what may be more important, also in their orientation to government and politics.

It will not do to erase from our collective memory the fact that caste quotas, like communal electorates, were inventions of colonial times. Our nationalist historians, of both the Marxist and the Gandhian persuasions, never tire of painting the colonial administration in the darkest possible colours. If, on the other hand, we believe that caste quotas are inherently just, we must give due credit to our colonial rulers for teaching us so much about them. I happen to believe that caste quotas are bad, and therefore must put them down on the debit side of colonial administration, without necessarily assigning evil intentions to it in every respect.

Caste quotas and communal electorates were parts of the same package of measures devised by the British for the governance of India. In fact, the term 'communal' had somewhat different shades of meaning in north and south India. In the north, it related above all to divisions on the basis of religion, particularly to the division between Hindus and Muslims. In the south, the same term, in such phrases as 'Communal Award' or 'Communal G. O.', referred primarily to divisions between castes or groups of castes, particularly the one between Brahmins and non-Brahmins. At least in the political arena, in north India, the division between Hindus and Muslims overshadowed to a large extent the ones between castes among the Hindus, whereas this did not happen in the south.

Caste quotas in public life were instituted in south India in the wake of the Montagu-Chelmsford Report of 1918, and became established as permanent features of government and politics in the succeeding decades. In Madras presidency, the Justice party set about distributing administrative appointments among the various castes and communities from 1921 onwards. These procedures were codified in the Communal G. O. of 1927 which laid down a roster system that operated till 1947 when it was revised. A similar policy was adopted in Mysore, following the recommendations of the committee of Sir Leslie Miller appointed in 1918.

When they were first introduced into south India in the 1920s, caste quotas were no doubt welcomed by some, but it is not true that they were welcomed by all. They were welcomed by the rising middle class among the non-Brahmins, supported overtly and covertly by British commercial and administrative interests. The opponents of caste quotas, representing the views of Gandhi, Rajaji, Satyamurti and others were characteristically dubbed by the former as an upper-caste oligarchy. The opposition between the two points of view was well represented by the two leading dailies of Madras: *The Hindu*, representing nationalist interests under upper-caste leadership, and *The Mail*, then representing British commercial interests and calling for safeguards against domination by the Brahmins.

Under colonial rule, positions of authority did not become available to Indians all at once, or under terms and conditions chosen by them. The terms were generally set by the colonial government. It is true that the Congress party tried to bargain for better terms, but its bargaining power was not unlimited, and there were other parties only too eager to set further limits to it: the Justice party in the south and the Muslim League in the north. The homology between the Justice party and the Muslim League comes out both in their opposition to the Congress and in their support of the Communal Award.

There was an additional reason why the opposition to reservations did not take the form of a mass movement when they were instituted in south India in the 1920s. Being a part of the non-Brahmin movement, they were designed primarily to restrict the predominance of the Brahmins in public life. The Brahmins occupied a unique position in south Indian, particularly Tamil, society. Their intellectual and cultural pre-eminence became even more pronounced as British rule became established. Yet, they were a very small section of society, less than 4 per cent of the population in Madras presidency. When the backlash came, they were neither equipped nor prepared to put up any effective resistance to it. They adapted themselves to the new conditions in various ways, including migration to the north where many of them did extremely well in the services of the union government in the early years of independence. They simply did not have the resources to organize a mass movement against caste quotas on their home ground.

When independence came, the Constituent Assembly sought to establish a new consensus in which the individual and the nation would prevail over caste and community. Its deliberations were held under the shadow of the partition of India, and enlightened opinion in most parts of India set itself resolutely against Communal Awards and Communal G. O.s. There can be no doubt that, as far as these matters went, a new spirit prevailed in the Constituent Assembly. That spirit lingered for some time, and has probably left a mark on many minds. It was that spirit that was expressed in the letter with which Kaka Kalelkar forwarded the Report of the first Backward Classes Commission to the President. He wrote, 'National solidarity demands that in a democratic set up Government recognize only two ends—the individual at one end and the nation as a whole at the other—and that nothing should be allowed to organize itself in between these two ends to the detriment of the freedom of the individual and the solidarity of the nation'.

In the south, which did not experience the trauma of partition in the same way as the north, the practices established by Communal Awards and Communal G. O.s continued, despite the changes introduced by the Constitution. There were no doubt changes in the form and extent of caste quotas, but these too were made, at least initially, as a result of the intervention of the Supreme Court, most notably in Champakam Dorairajan's case, and again in Balaji's case. The caste quotas that were introduced by the British in the teeth of nationalist opposition have now become a way of life in south India. One may say that it is a good way of life, depending on taste, judgement or acumen. But those who are aware of the historical circumstances under which that way of life was introduced and institutionalized, will find it hard to believe that it can be defended by an appeal to Gandhian principles.

The social and political climate in north India, and indeed in the whole of India, in 1990 is very different from what it was in Madras presidency in 1920. Opposition to any move, whether good or bad, by the government can now be organized much more swiftly and extensively than was conceivable seventy years ago. That, however, is only one side of the matter, and perhaps not the most important one. Indeed what has struck many observers about the current movement against the recommendations of the Mandal Commission is the absence of political organization in it and the indifference of all political parties to it.

What has sustained the movement and given it its peculiar intensity is not any kind of political organization, but a certain sense of moral outrage. It is this sense of moral outrage and not any political force that has unsettled every political party. It is true that the British policy of pitting caste against caste, and community against community in the name of justice and fairplay aroused widespread resentment and hostility, but it did not create the kind of response that has now come to the surface. Perhaps Indians of an earlier generation could never feel towards their alien rulers the sense of outrage that their descendants now feel towards the leaders they have themselves freely chosen.

In conclusion, there is more than one lesson to be learnt from the south Indian experience. Some say that since caste quotas have worked in the south, there is no reason why they cannot be made to work throughout the country. That is certainly a matter for reflection, but our reflections cannot stop there. To me the most important lesson of the south Indian experience relates not to the workability of caste quotas but to their irremovability. Nobody who knows the history of caste quotas in that part of the country can seriously maintain that they can hope to see them removed in their own lifetime. Therefore, it is a little disingenuous to hold the south Indian case as an example and also argue that caste quotas are, after all, a transitional phenomenon that can be dispensed with after they have been made to work for ten or fifteen years. In the south, they have been at work for seventy years, and all the evidence indicates that they will continue to be at work for at least another seventy years.

VII

The Millstone of Reservations: Liberalization with Caste Quotas

Whatever gains the new policy of economic liberalization might have secured for the country will be wiped out by the policy of massive caste quotas in employment. If there are thinking persons in the government, they must surely realize that the principle of liberalization in the economy is directly opposed to the principle of quotas in public institutions. There is no way in which the two domains can be effectively insulated from each other in our kind of society. Sooner or later they will be bound to collide, and the result of the collision will be neither economic efficiency nor social justice.

Forty years ago the Indian intelligentsia started with a certain

The Times of India, 12 October 1991.

faith in the capacity of the state to play the leading part in the transformation of society and culture. A great many things needed to change, and it was natural to believe that a national government would do all the things that its colonial predecessor had either failed or refused to do. It will be difficult to appreciate the faith reposed by the Indian intelligentsia in the government if we fail to keep in mind the auspices under which it assumed office at the time of independence. In the years preceding independence, the landlords and the capitalists had acted badly by the people of India; those who took charge of the new government had, as leaders of the nationalist movement, acted well by them.

This faith in the capacity of the government to change all things for the better found its characteristic expression in the idea of centralized planning. That idea reached well beyond the limits of a technical economic exercise; it came to signify a whole philosophy. Without much experience of the practical limitations of economic planning, it was not too difficult to believe that it could be used not only to raise the national product or lower unemployment, but also to establish equality and social justice. In the fifties and sixties the Indian intelligentsia placed much faith in economic rationality, but their faith was in the rationality of the plan as against that of the market.

By the eighties, the limits of a centrally-regulated economy had begun to manifest themselves. Knowledgeable persons, including some civil servants, were saying that the government had landed the economy in a mess. But established habits of mind do not change very easily or all at once. The responsibility for the failure on the economic front was assigned first to individual politicians and bureaucrats and then to the whole political class. It was not immediately apparent that the fault lay deeper in the fact that the government had, in the name of socialism, vastly overextended its activities. Thus, even those who became progressively disenchanted with each successive plan often felt that things could be set right with bigger plans and more effective regulation.

But the climate of thinking in the world as a whole has now become more favourable towards liberal economic systems. In one country after another, including China and the U.S.S.R., the negative economic consequences of always promoting the bureaucracy at the expense of the market have been dramatically brought to light. In eastern Europe, the accumulated resentment against the bureaucracy for its mismanagement of the economy has created the danger that a blind faith in the state may be replaced by an equally blind faith in the market.

Fortunately, in India the excesses of the state in the name of socialism have been mild in comparison with the U.S.S.R. and other east European countries. The question here is not of doing away with the role of the government in the economy but rather of limiting it to those areas in which it can be most effective. Even the strongest advocates of liberalization or a market-friendly approach have not called for an end of either planning or the public sector. What they want is more freedom for economic activities to establish themselves in their own domain and to operate according to their own logic.

It is in this context that we have to appreciate the relief with which the new package of economic policies introduced by the government since July has been received both within and outside India. On the whole, responsible persons have expressed misgiving rather than hostility towards what the government is seeking to achieve on the economic front. There are misgivings on many counts, and these need to be carefully examined.

The success of the new economic policy will depend on a number of factors apart from the economic ones in the narrow sense of the term. Above all, it will depend on a new set of attitudes and orientations and a new framework of roles and relationships. What has been lacking in our economy so far is dynamism, and it cannot be too strongly emphasized that economic dynamism is more a matter of attitudes, values and institutions than of material resources. Forty years of political and bureaucratic patronage have stifled and suffocated all those qualities that stand for economic dynamism in human terms. The real obstacles that the new economic policy will encounter will be from ingrained attitudes and relationships that are inimical to enterprise, initiative and flexibility.

The social and psychological preconditions for a turn around in the economy are probably well understood in the ministry of finance. All the talk there has been about initiative and enterprise, about allowing the best talent to make room for itself in every field through open competition. At the same time, the government's policy on caste quotas in public employment is bound to choke the very processes on which the new turn in the economy is vitally dependent. It hardly needs to be emphasized that recruitment through caste quotas is the very antithesis of selection on the basis of success in open competition. From where will the much needed dynamism in the economy come if employment remains entangled in caste quotas?

The economic ministries have at last begun to view quotas with misgiving and suspicion, whereas thirty or forty years ago one only spoke against them in a weak voice. Liberal economic doctrine views quotas as obstacles to the smooth functioning of the market and requires that they be reduced to the minimum. If quotas are bad in general, quotas based on caste and community are particularly retrograde. They not only stifle initiative but make factional strife endemic. As caste quotas come to be extended indefinitely, they turn every organization from a work site into a bargaining centre. It is characteristic of the politics of reservation that, whereas Mr V. P. Singh's government wanted quotas to the extent of 49.5 per cent, the present government has offered to increase them by another 10 per cent; the introduction of some economic restrictions does not substantially alter the fact that not merit but some other criteria will be the principal basis of recruitment for reserved positions. An economy cannot be made competitive by persons who owe their positions in an organization to birth rather than achievement.

It can of course be said that economic efficiency will be looked after by the private sector while the public sector attends to social justice. Indeed, so intense has been the reaction against corruption and inefficiency among politicians and civil servants that there are now some who are prepared to write off the public sector and to get on with their business. Educated Indians have begun to prepare themselves mentally for caste quotas in the civil service, not because they believe that caste quotas are good, but because they think that the civil service is so bad that it is almost beyond recovery. Those who see the bureaucracy as a source of rental income are likely to agree that the rents should be shared more widely among the different castes and communities.

But such a view of the relationship between the private and the public sectors—or between enterprise and governance—would be short-sighted. No matter how much we welcome the opening up of the economy, we cannot do without a public sector in India. The public sector will continue to occupy a significant space in the Indian economy, and it is important to ensure that it does not become choked by an employment policy that is inherently defective. A public sector in which employment is governed by caste quotas will sooner or later infect the whole economy with its own inefficiency and corruption.

Nor is it a question of the public sector in only the narrow economic sense of the term, but of public institutions in the widest sense. Caste quotas threaten the health of a whole array of administrative, educational, scientific, legal, financial and other institutions in India. These institutions have a vital role to play in the modernization of Indian society without which no programme of economic liberalization can prosper. We have learnt a few hard lessons through our failures on the economic front. We cannot benefit from those lessons if we expect the market alone to deliver all the goods, while allowing our public institutions to decay.

VIII

The Politics of Caste: Competition in Backwardness

In the four southern states of Andhra Pradesh, Kerala, Madras and Mysore, a large number of castes are entitled to certain important privileges by virtue of their being grouped together as Backward Classes. Mysore state provides the most extreme example with well over 50 per cent of its population listed as Backward in addition to the Scheduled Castes and Tribes, comprising about 14 per cent of the total population. The Backward Classes in Mysore state today include the two most powerful castes which between themselves dominate the state legislature as well as other important political bodies.

In the centre there appears to be some misgiving about the advisability of continuing the use of caste criteria in the matter of

The Indian Express, 22 and 23 August 1961.

reservations either for government jobs or for admission to technical institutions. This seems to contrast sharply with the prevalent mood in at least some of the states. The Final Report of the Mysore Backward Classes Committee, popularly called the Nagan Gowda Committee Report, published in June 1961, has once again reinstated caste as the basis for according preferential treatment. The reactions which followed the publication of the Report in Mysore state are as important indicators to the involvement of caste in politics as the recommendations themselves. The Report had originally excluded the powerful Lingayat caste from the category of Backward Classes. This was immediately followed by intense political campaigning by the Lingayats, leading within a very short time to their also being included as Backward.

It is now more or less well known that claims to Backward status have in practice got to be assessed in terms of the political strength of the claimants, apart from other considerations. This will continue to be the case so long as people are classified as Backward or Forward on the basis of caste. The role of caste in political affairs has grown from strength to strength, and a policy on the part of the government to give protection on the basis of caste can hardly be expected to reverse the current.

One of the consequences of the introduction of the parliamentary form of government in India was the enormous scope it provided for castes to organize themselves politically. In this regard, some castes have been more successful than others, and in several states in India, political affairs are controlled by two or three dominant castes. In Mysore state, as well as elsewhere, the important non-Brahmin castes developed the technique of capturing government jobs and securing other important concessions on the plea that they were Backward. Because these advantages had to be secured through parliamentary means, the leading dominant castes set about organizing themselves as successful pressure groups. The concessions thus gained in their turn helped the castes which were numerically stronger and better organized to enhance their political power. The net result of forty years of political manoeuvering has been that politics in Mysore state is today largely controlled by the two dominant castes, the Lingayats and the Okkaligas. These castes owe their political success in no small measure to the numerous governmental concessions gained on grounds of being Backward.

Reservation of seats on the basis of caste, it is easy to see, is likely to harden the structure of caste instead of reducing the disparities between the different sections of the population. Those who advocate the use of caste criteria maintain that such reservations are necessary from the viewpoint of 'social justice', since wide disparities exist between different castes. It is important to note, however, the active part played by the Backward castes themselves to force their way into the category of Backward through the use of various kinds of pressure. The agitation which was set afoot by the Lingayats on their being excluded from the Backward Classes and their subsequent inclusion bear witness to the role of caste in the politics of Backwardness.

In Mysore state the situation has acquired added piquancy owing to the fact that the two castes which control the politics of the state both claim to be Backward; for Lingayats to be excluded from the category of Backward would mean political defeat at the hands of their chief rivals, the Okkaligas. Claims to Backwardness and political power reinforce each other in Mysore state today. The Lingayats owe their political power in part at least to the concessions they have acquired as a Backward community; they can continue to enjoy the concessions accorded to Backward communities because of their political power.

In this context one has to distinguish between the Scheduled Castes and Tribes on the one hand and the Backward Classes on the other. The latter are not only economically and socially better off, but, what is more important, are often in a position of great political strength. The decision to accord preferential treatment to the Scheduled Castes and Tribes was perhaps justified in the circumstances in which it was taken. The communities constituted minorities which had suffered social oppression for generations and were really at the bottom of the social and economic ladder. The motives behind their upliftment were largely humanitarian. The measure taken were justified, at least for the time being, because of the large and glaring gaps between these and the 'forward' communities.

The concessions progressively granted to a wide variety of castes by labelling them as Backward cannot, however, be always explained in purely humanitarian terms. There can be little doubt that in at least a number of cases, political pressures have been equally important, if not more so. The recent incidents in Mysore state referred to above, show how the Lingayats, a very powerfully organized caste, have been able to force their way into the privileged category of Backward Classes.

It is important to take account of the difference in attitude between the centre and the states with regard to the use of caste criteria. A dominant caste such as the Lingayat or the Okkaliga has limited extension, and operates most successfully at the level of state politics. The centre has at least no direct interest in giving political backing to particular castes at the state level. But although the centre is more free from caste politics, leaders of the state legislature as well as the state Congress party can exert pressure on the centre for the continuance of protection on the basis of caste. Unless the centre is bold enough to break through the caste barrier and to reject all schemes of preferential treatment on a communal basis, state politics is bound to get more and more involved with caste.

It is interesting to note that whereas the problem of deciding which communities should be listed under Scheduled Castes and Tribes is handled by the centre, matters relating to the Backward Classes are left in the hands of the states. Perhaps it is realized that the Scheduled Castes and Tribes are not strong enough to hold their own ground in state politics. The Backward castes in Mysore, or at least the leading ones among them, are more than strong enough to hold their own. Perhaps their very strength makes the centre a little hesitant to pursue a firm line regarding the abolition of privileges based upon caste.

The use of caste criteria in the matter of reservation is both a cause and a consequence of the growing influence of caste in politics. Mysore state faces the paradox of having to grant protection to more than 80 per cent of its total population on the ground that they are either Backward or Scheduled.

What is to be the position of 'Forward' communities in the new scheme of things? They have to suffer the political insecurity that is the lot of all minorities. In a demoratic system of government, numerical weakness is a severe handicap, just as numerical strength is a sure source of power. This can be well understood by comparing the political fortunes of the Brahmins in Mysore state with those of two numerically powerful Backward castes, the Lingayats and the Okkaligas. The latter have grown continuously in strength until now they virtually control the politics of Mysore state. The former have been reduced to a position where many of them feel that their only hope lies outside the state.

In deciding whether a community is Backward or not, the Nagan Gowda Committee have taken a number of factors into account. These are: position in the traditional hierarchy, level of education, level of income, etc. Perhaps it has not struck the Committee that the political strength or weakness of a caste may also be a consideration for protection. One wonders how much it matters whether a politically dominant caste such as the Okkaligas are at the middle or the bottom of the traditional hierarchy. On the other hand, the Brahmins, in spite of their superior position in the traditional hieararchy and their higher average level of education, can justly claim protection on the grounds of political insecurity.

The whole argument shows the futility and the danger behind the use of caste criteria in the matter of government protection. Caste cannot and should not be the guiding priciple in this matter because of its invlovement in politics. To legitimize protection on the basis of caste can only lead to an increase in its hold over politics. If protection has to be provided in any case, it is imperative that more rational criteria be devised.

To continue to give protection on the basis of caste can lead only to a perpetuation of the pressure groups which have already become an important feature of state politics. Dominant castes have a vested interest in continuing to be labelled as Backward. Now that they have political power, they can act in a more concerted manner to have their interest safeguarded. This is precisely what the Lingayats and the Okkaligas have been doing in Mysore state. Examples may be provided from other states as well.

If, on the other hand, protection is given, not to communities but to individuals, on the basis of income, occupation or some similar criterion, the situation will be different. This is bound to go a long way in minimizing the role of pressure groups, many of which are based upon caste. It does not require much foresight to realize that a category of individuals belonging, say, to a particular income group, has much less scope for organizing itself as a pressure group than has a powerful dominant caste. Apart from lacking the political organization of a corporate community, such a category is by its very nature a more fluid and flexible entity.

Apart from the obvious political dangers inherent in the use of caste as the basis for protection, there are other factors which have to be taken into account. For one thing, the reservation of posts on the basis of caste is bound to result in a lowering of standards. This fact has already made itself evident in academic life in Mysore state as well as in other places. When a person is more or less assured of a position because he belongs to a particular caste, he is naturally inclined to put in less than his best. A candidate who is more able and efficient is often bypassed because he does not belong to the right caste. This is not only against the principle of social justice, but it also leads to a continuous decline in standards.

It has also to be borne in mind that communities which are listed as Backward or More Backward are often fairly large and heterogeneous. Not all the members of such a community deserve the patronage they are officially entitled to receive. Some of them are indeed men of wealth, social influence and political power. And these are precisely the people who know best how to take advantage of the benefits conferred on their community as a whole by the government. It is a matter of common knowledge that among the Backward Classes the benefits of governmental protection are syphoned off to the dominant castes and their leaders. Thus, the very purpose of granting protection is defeated.

Many, if not most, of these difficulties can be eliminated by using a set of rational criteria in the definition of Backward categories. Income, occupation, literacy, education, rural as opposed to urban residence may be counted as some of these criteria.

It is a bit of a paradox that a government which is prepared to ignore caste to the extent of omitting any reference to it in the census, has to come back to it in a matter of such vital importance. The only way to break through the edifice of caste is to take account of its tremendous potentialities in a political system of the present type. Educated people in India are often innocent about caste because many of the ritual taboos are breaking down and because in large cities like Bombay and Delhi there are occasional instances of inter-caste marriages. It is strange how these things have rendered the Indian intelligentsia blind to the increasing influence of caste in politics. In the present political set-up, today as never before, caste has the possibility of providing the severest challenge to the integrity of India as a nation.